T0327530

67th Conference on
Glass Problems

67th Conference on Glass Problems

*A Collection of Papers Presented at the
67th Conference on Glass Problems,
The Ohio State University, Columbus, Ohio,
October 31–November 1, 2006*

Editor
Charles H. Drummond, III

WILEY-INTERSCIENCE

A John Wiley & Sons, Inc., Publication

Published by John Wiley & Sons, Inc., Hoboken, New Jersey.
Published simultaneously in Canada.

For general information on our other products and services or for technical support, please contact our Customer Care Department within the United States at (800) 762-2974, outside the United States at (317) 572-3993 or fax (317) 572-4002.

Wiley also publishes its books in a variety of electronic formats. Some content that appears in print may not be available in electronic format. For information about Wiley products, visit our web site at www.wiley.com.

Wiley Bicentennial Logo: Richard J. Pacifico

Library of Congress Cataloging-in-Publication Data is available.

ISBN 978-0-470-19065-4
ISBN 978-0-470-19066-1 (special edition)

10 9 8 7 6 5 4 3 2 1

Contents

ENVIRONMENT

ENERGY

MELTING AND REFRACTORIES

Foreword

The 67th Conference on Glass Problems was sponsored by the Departments of Materials Science and Engineering at The Ohio State University.

The director of the conference was Charles H. Drummond, III, Associate Professor, Department of Materials Science and Engineering, The Ohio State University.

Dean William A. Baeslack, College of Engineering, The Ohio State University, gave the welcoming address. Former Chairman John Morral, Department of Materials Science and Engineering, The Ohio State University, gave the Departmental welcome.

The themes and chairs of the five half-day sessions were as follows:

The Future of the Glass Industry
Gerald DiGiampaolo, PPG Industries, Pittsburgh PA
John Tracey, North American Refractories, Cincinnati OH

Safety
Terry Berg, CertainTeed, Athens GA
Larry McCloskey, Toledo Engineering, Toledo OH

Modeling
Ruud Beerkens, TNO Glass Technology – Glass Group, Eindhoven, The Netherlands
Elmer Sperry, Libbey Glass, Toledo OH

Environment
Tom Dankert, Owens-Illinois, Toledo OH
Dick Bennett, Johns Manville, Littleton CO

Energy
Bob Thomas, Corning, Corning NY
Sho Kobayashi, Praxair, Danbury CT

Melting and Refractories
C. Philip Ross, Glass Industry Consulting, Laguna Niguel CA
Dave Watters, Osram Sylvania, Exeter NH

Preface

In the tradition of previous conferences, started in 1934 at the University of Illinois, the papers presented at the 67th Annual Conference on Glass Problems have been collected and published as the 2006 edition of The Collected Papers.

The manuscripts are reproduced as furnished by the authors, but were reviewed prior to presentation by the respective session chairs. Their assistance is greatly appreciated. The Ohio State University is not responsible for the statements and opinions expressed in this publication.

CHARLES H. DRUMMOND, III
The Ohio State University
Columbus, OH

Acknowledgments

It is a pleasure to acknowledge the assistance and advice provided by the members of Program Advisory Committee in reviewing the presentations and the planning of the program:

Ruud G. C. Beerkens—TNO-TPD

Dick Bennett—Johns Manville

Terry Berg—CertainTeed

Tom Dankert—Owens-Illinois

Gerald DiGiampaolo—PPG Industries

H. "Sho" Kobayashi—Praxair

Larry McCloskey—Toledo Engineering

C. Philip Ross—Glass Industry Consulting

Elmer Sperry—Libbey Glass

Robert Thomas—Corning

John Tracey—North American Refractories

David A. Watters—Osram Sylvania

MARKET TRENDS

CONTAINER GLASS UPDATE

Rick Bayer
Glass Packaging Institute

ABSTRACT

The Glass Packaging Institute (GPI) is the trade association representing the North American glass container industry. Through GPI, glass container manufacturers speak with one voice to advocate industry standards, promote sound environmental policies and educate packaging professionals. GPI member companies manufacture glass containers for food, beverage, cosmetic and many other products. GPI also has associate members that represent a broad range of suppliers.

GLASS CONTAINER MARKET

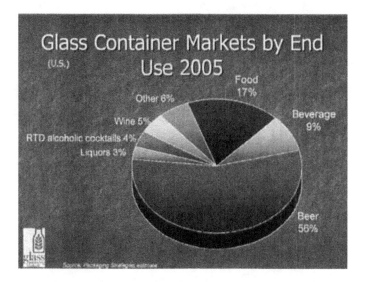

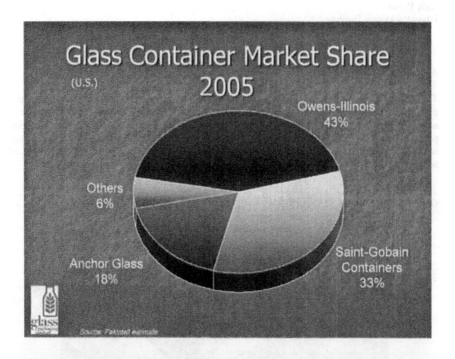

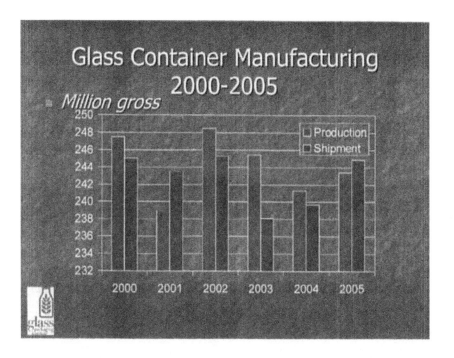

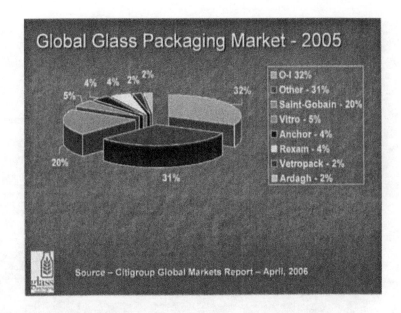

ENVIRONMENT AND ENERGY

Environmental Activities

- OZONE Transport Commission
    - Requesting dramatic reductions in NOx
- California Greenhouse Gas Emissions Law
    - Goal of reducing all components back to 1990 levels
- EPA Continues Heavy Enforcement Activity
    - Permitting pressure versus compliance
    - Capital and operating costs of abatement solutions
    - Increasing complexity of reporting and recordkeeping requirements

Energy Issues for Containers

- Increased energy costs
- Effect on industry margins worldwide
- No major relief envisioned at this time

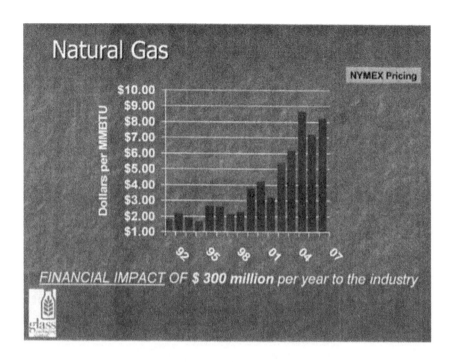

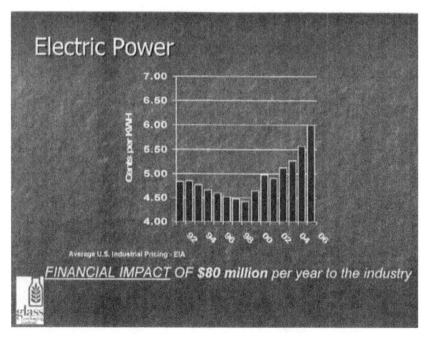

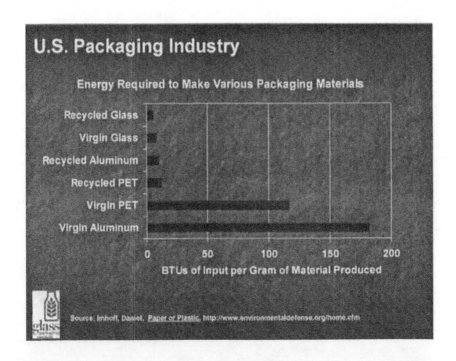

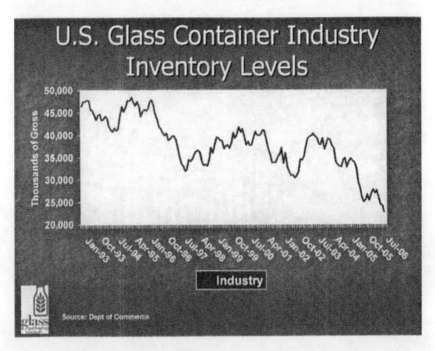

INDUSTRY TRENDS

Demand for Glass Containers
- Main driver is substitution / consumer preferences
- Beer, wine & liquor and cosmetics remain as the core / most resilient segments in glass
- CSD fully converted, food penetration already high in main segments, beer unlikely to move

Sustained Growth for the UK Market is Predicted*
- Market Development Report predicts glass containers will grow 8% by 2010 in UK Market
- Demand from premium foods will result in 9% growth between 2005-2010 for colorless food and drink containers

*Source – UK Glass Packaging Market Development Report; February, 2006

Mexican Beer Market
- Mexican beer market has grown consistently at ~6% per year
- Consumption per capita ~50 liters per year
- Most popular form of packaging is the glass returnable bottle with 78% share

Key Growth Markets for Glass
- Main Countries: Vietnam, Thailand, China, and the Philippines
- RTD/FABS – offering growth opportunities in selected market
- Beer – growth unabated in certain volume   market (China), strong growth in single serve long neck beer for the premium end beer market
- Non alcoholic beverage – renewed interest by multinational for glass in CSD
- Food – exported food in jars to Europe

PROMOTING GLASS PACKAGING

Outreach to Packaging Decision-Makers

Marketing Programs Include:
- Trade Magazine Advertising
- Clear Choice Awards
- Article Placements in Trade Publications
- Website/E-Newsletter

Glass continues to promote a premium presence across many market segments

GLASS PACKAGING TRENDS

- Organic Growth Pushing Demand for Glass Containers
- Glass Packaging for Branding and Premium Positioning
- All-natural Products Using Glass to Communicate Purity
- Commitment to Recyclable Packaging and Sustainability

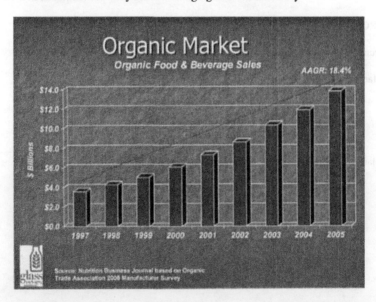

GPI OUTREACH

Earned Media Outreach
- "New Garden Select is Clearly Better" *Food & Drug Packaging* (January 2006)
- "Going First Class with Glass" *Food Processing* (March 2006)
- "Make it Better, Not Cheaper" *Food Engineering* (March 2006)

Competitive Intelligence Alert
- Competitive packaging news and trends
- Packaging driven by convenience news
- Global consumer trends that influence packaging

GPI E-Newsletter
- New product introductions and conversions to glass
- "A Day in the Life" (Profile packaging decision-makers)
- Organic food trends

GPI Website (www.gpi.org)
- "Located a Supplier" navigation bar
- Real time news and updates
- New products introduced in glass updated monthly
- GPI program information and educational resources

Survey Shows Preference for Glass*
- 82% feel that glass does the best job of ensuring quality and healthiness of food
- 78% of consumers choose glass packaging for maintaining purity of product
- 90% of those surveyed prefer glass packaging for baby food
- 96% prefer glass packaging for wine
- Study showed that 90% of respondents who drink beer believe it tastes better when served in a glass bottle
- Only 2% of that same group believed beer tastes better when served in an aluminum can
- In a recent survey, 96% of respondents stated they prefer wine packaged in glass
- Only 2% of respondents prefer their wine packaged in paper (Tetra Pak)

*Sources – Newton Research - April, 2006 and Dragon Rouge Survey - May, 2006

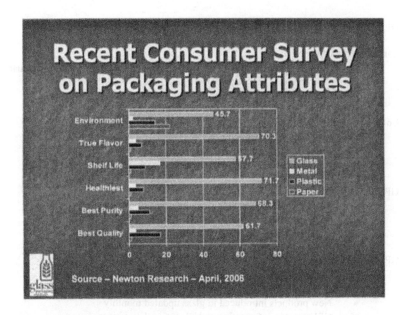

Academic/Packaging Education Program
- Lecture Series
- Plant Tours
- Interactive CD-ROM
  - 1,200+ CDs Distributed To The Academic Community

Topics for Lecture Program
- Surface Treatment
- Light Weighting
- Retail Trends
- Labeling & Decorating Techniques
- Design Criteria
- Recycling
- Glass Packaging Fundamentals

For more information, please visit www.gpi.org

SAFETY

SAFETY

SAFETY IN CONSTRUCTION

Laura Gray
Operations Manager
Labor Alliance Group, Inc.

ABSTRACT

According to the Bureau of Labor Statistics there were 1,186 fatalities in the construction industry in 2005. Falls lead the way with 394 deaths. Transportation was next with 312 followed by contact with objects and equipment at 245 deaths. Exposure to harmful substances or environments and fires and explosions had 202 fatalities and assaults and violent acts had 32 deaths. The good news is that this is a 4% decrease from 2004. Safety in the construction industry has unique requirements. Not only do we have to comply with the OSHA 1926 CFR (Code of Federal Regulations) for the Construction Industry but also have to have knowledge and understanding of the OSHA 1910 CFR for General Industry.

The workforce on construction projects is different for each project and the project itself has different needs. A challenge to the employer is you don't know what type of safety training the workers you are hiring have had or the level of training. It can be assumed all will need additional training for the project. If you put the cost of the time for training in the bid, on the assumption you have to provide training for everyone you are hiring, you might price yourself out of the job. If you don't put anything in for training and find you have to provide site specific safety training it can be expensive when you start adding up the man hours spent in training. This can put a huge dent in your profit margin, especially when you start purchasing safety supplies and equipment to accompany the training.

A good example of unexpected training costs would be to observe workers wearing their fall protection harnesses incorrectly. They do not have them properly fitted to their bodies which, if they would fall, could cause more severe injuries. This observation tells you they do not have proper training and it is your responsibility to provide the necessary training to ensure their safety (See CFR 1926.21, Safety Training and Education).

There is no easy answer. When it comes to construction projects, a proactive approach to safety and a good safety officer with communication skills is an excellent start. We have a good understanding of the safety requirements that must be provided for the workforce. A lot of time and money is spent on educating and training workers regarding safe work practices.

Figure 2 is the face of safety today. But is it enough? Do we need more than just personal protective equipment?

LABOR STATISTICS

The Department of Labor keeps statistics on workplace accidents. Although statistics can be boring, without them we would not know where to look to find trends and areas that are in need of attention for the prevention of accidents. Statistics also give insurance companies the basis for calculating our insurance premiums and are the basis for experience ratings for workers compensation insurance. Increases in these costs make it necessary to eliminate jobsite accidents.

Every year OSHA compiles statistics on violations to its standards. During 2005, there were 105,817 violations to standards ranging across all industrial segments under federal OSHA jurisdiction, with adjusted penalties of nearly $34 million. Because of the importance of statistics in our industry I have included last years top twenty-five violations and the top ten most frequently sited violations.

The following table lists the top 25 most frequently violated regulations for Part 1910 General Industry. These are federal statistics and are the result of violations that occurred in states that are regulated by federal OSHA. Citations for states that operate their own occupational safety and health programs are not included in these statistics.

Table I. Top 25 General Industry Violations by Subparagraph (January 1, 2005 through December 31, 2005)

	Subject	OSHA Standard	Total Violations	$ Initial Penalty	$ Adjusted Penalty
1	Hazard communication-Written program	1910.1200(e)(1)	2,395	$1,110,276	$427,951
2	Machine guarding-Types of guarding	1910.212(a)(1)	1,587	3,066,794	972,789
3	Hazard communication-Employer must provide hazard information and training	1910.1200(h)(1)	1,124	256,272	90,687
4	Machine guarding-Point of operation guarding	1910.212(a)(3)(ii)	811	1,869,156	704,841
5	First aid-Eye wash/emergency shower facilities not in near proximity to employees	1910.151(c)	784	904,354	345,566
6	Guarding floor openings, platforms and runways	1910.23(c)(1)	744	1,303,417	401,865
7	Lockout/tagout-Establish an energy control program	1910.147(c)(1)	732	842,465	286,927
8	Abrasive wheel machinery-Exposure adjustment/safety guards	1910.215(b)(9)	729	285,041	99,502
9	Hazard Communication-MSDS available for each hazardous chemical	1910.1200(g)(1)	718	98,975	37,093
10	Lockout/tagout-Written energy control procedures	1910.147(c)(4)(i)	700	1,203,548	393,020
11	Respiratory protection-Provide medical evaluation prior to fit test and respirator use	1910.134(e)(1)	694	467,800	75,976
12	Respiratory protection-Establish a written program	1910.134(c)(1)	691	472,093	159,669
13	Electric-Wiring methods, components and equipment-Cabinets, boxes/Conductors	1910.305(b)(1)	662	483,650	153,245
14	Hazard Communication-Employ information and training	1910.1200(h)	604	135,350	57,565
15	Powered industrial trucks-Operator training	1910.178(l)(1)(i)	563	752,932	194,475
16	Hazard communication-Labeling containers	1910.1200(f)(5)(i)	547	180,140	73,268
17	Mechanical power transmission-Pulley guarding	1910.219(d)(1)	544	550,791	173,771
18	Electric-Listed and labeled equipment must be used or installed according to instructions	1910.303(b)(2)	515	402,545	125,915
19	Lockout/tagout-Training and communication	1910.147(c)(7)(i)	512	535,450	205,612
20	Abrasive wheel machinery-Must use work rests	1910.215(a)(4)	502	350,752	119,627
21	Electric-Wiring methods, components and equipment-Cabinets, boxes/Covers	1910.305(b)(2)	501	342,162	93,747
22	Compressed air-Reduce to less than 30 psi	1910.242(b)	494	366,789	130,136

23	Lockout/tagout-Annual procedure inspection	1910.147(c)(6)(i)	493	370,575	121,924
24	Personal protective equipment-Provide, use, maintain equipment in a sanitary and reliable condition	1910.132(a)	479	840,540	238,243
25	Hazard communication-Maintain MSDS	1910.1200(g)(8)	468	66,616	16,901

While the following citations listed in Table II are not part of the 1910 General Industry regulations, they apply to all employers and also rank among the top violators for all industry groups under Federal OSHA jurisdiction.

Table II

	Subject	OSHA Standard	Total Violations	$ Initial Penalty	$ Adjusted Penalty
1	OSHA Act-General Duty Clause-Employer must provide a safe work place for all employees	5(a)(1)	1,177	3,711,904	977,212
2	Record work-related injuries and illnesses on OSHA 300 log, 301 incident report, and 300A summary	1904	2,037	1,135,920	351,979
3	Inspections, citations, and proposed penalties	1903	229	91,230	6,226

Table III - OSHA's 10 Most Frequently Cited Standards of 2005

	Violation	OSHA Standard	Total Violations
1	Scaffolding Compliance Training: Construction sites	1926.451	8,891
2	Hazard Communication Compliance Training: Hazard Communication	1910.1200	7,267
3	Fall Protection Compliance Training: Construction Sites		6,122
4	Respiratory Protection On-line Training Available: Respiratory Protection	1910.134	4,278
5	Lockout/Tagout Compliance Training: Lockout/Tagout	1910.147	4,051
6	Powered Industrial Trucks Compliance Training: Powered Industrial Trucks	1910.178	3,115
7	Electrical Wiring Compliance Training: Electrical Safety	1910.305	3,077
8	Machine Guarding Compliance Training: Machine Guarding	1910.212	2,956
9	Electrical General Requirements Compliance Training: Electrical Safety	1910.303	2,348
10	Ladders Compliance Training: Construction sites	1926.1053	2,276

CONSTRUCTION SITE ACCIDENT FACTS

The following information is compiled from the Bureau of Labor Statistics, United States Department of Labor and from the book Accident Facts, published by the National Safety Council (1996 Edition).

- Accidental injuries are the fifth leading cause of death in the United States.
- In 2005 incidence rates in the construction industry was 6.3 cases per 100 full time employees.
- One out of every 10 construction worker is accidentally injured every year.
- The most common accident at construction sites is falls, either on the same level or from height. More fatalities occur from falls than any other construction activity.
- Ironworkers are the construction trade with the greatest likelihood of being injured on a construction site.
- More ironworkers are injured during decking operations than any other ironworking activity.
- The back is the part of the body most frequently injured at work accounting for nearly 25% of all work-related injuries. Injuries to the legs, arms and hands are the next most frequent bodily injuries.
- By percentage of population, there are fewer accidental deaths in New York that any other state and in New England than any other part of the country.
- The states with the greatest incidents of accidental deaths in the construction field are Texas - 134, Florida - 109, California - 102, Georgia - 55, Virginia – 50, Pennsylvania – 36, S. Carolina – 33, Illinois – 30. Ohio had 26 and Michigan had 20. The states with no construction fatalities were Rhode Island and Vermont.
- Motor vehicles are the cause of nearly 50% of all accidental deaths in the United States.

TRAINING

As you may have noted in the above statistics, lack of training is an issue. It can not be stressed enough how important training is for everyone's safety.

Where to start all the training that might be needed for each project is in the jobsite evaluation which should be done before the start of any project. Safety requirements can be determined at that time and plans for specific safety training. i.e., fall protection planning, respiratory requirements, etc. can be organized.

During hiring orientation a lot of safety information is given. It is important to make sure the employees understand what they have been told. The best way to do this is with a test at the end of the orientation that will show comprehension. The next way to provide on the job training is through tool box safety talks. Tool box safety talks are held during the shift. These are intended to familiarize workers with a variety of different safety topics, provide them with additional information and as a reminder to think safety. These talks can be done as often as needed and it is a good way to keep safety in the forefront of everyone's consciousness. Toolbox talks are inexpensive with time being the cost factor. We can zero in on subjects we cannot put in the hiring process because of the time limitations. Tools box talks can be anywhere from five to fifteen minutes depending on the topic. With these talks we can cover in more detail ladder safety, electricity and other site specific topics.

DIRECT AND INDIRECT COSTS OF ACCIDENTS

Every accident results in costs. These costs are either direct or indirect. Direct costs are reimbursed by your insurance company. Indirect costs are not.

1. Direct Costs: These are medical expenses incurred from the injuries in the accident and compensation.
2. Indirect Costs: These are non-billable costs. These include time lost from work; loss of earning power; economic loss to the injured family; lost time by fellow worker; loss of efficiency by the crew; lost time by the supervisor; cost of hiring a new worker; damage to tools and equipment; spoiled product; loss of production and overhead costs.

When an employee is injured or worse the hidden costs start creeping into view. Employees stand around talking about the incident. The would-haves and should-haves are being discussed. Next some of these employees will need to be interviewed to tell what they observed. Lack of production is becoming evident. The supervisor has to take time from his duties to interview witnesses and put it in an accident review report. The picture is becoming clearer that you are loosing time and production. What about the injured employee and his family? Their losses are just as real as the employer's.

Like an iceberg, the hidden costs of accidents are not readily seen but they are there, just below the surface and they are formidable. To give you an idea of just how much it can cost for an injury, below are calculations you can use to determine how much the indirect costs are.

Direct Cost

Total value of the insurance claim for the injury	$1,000.00

Indirect Cost

To calculate the indirect cost of the injury, multiply the direct cost by the cost multiplier. This depends on the size of the direct cost.

If your direct cost is:	Use this cost multiplier
$0 -$2,999	4.5
3,000 - 4,999	1.6
5,000 - 9,999	1.2
10,000 or more	1.1

Direct Cost	X	Cost Multiplier	=	Indirect Cost
$1,000	X	4.5	=	$4,500

Total Cost

Direct Cost	+	Indirect Cost	=	Total Cost
$1,000	+	$4,500	=	$5,500

You can now see that spending a few dollars for the proper tools and equipment can save you money. Because these costs are hidden, we don't think about them until they are uncovered after an accident.

OSHA has a computer program that is free of charge to all employers. This program is provided to assist you in determining how much accidents are costing your company. (Website address: http://www.osha.gov/dts/osta/oshasoft/index.html).

HUMAN ERROR IN OCCUPATIONAL ACCIDENTS

Human error plays a big role in accidents. "To err is human" applies to all of us. None of us are immune to this. But what causes us to commit errors? Not enough sleep, not enough training for the job at hand, improper tools, the mind preoccupied with home/social life, and job pressures. How many times have you stayed out late on a work night enjoying the company of your friends and family? Sometimes we go to work and are not as sharp and alert as we should be because our other lives interfere with our work life. On the other hand, most workers go to their jobs day in and day out without experiencing an adverse affects of going to work not fully rested and alert. Repeatedly doing this puts them in the mindset of "What, me worry?" They have a false sense of security since they have been doing it like this for so many years that an accident won't happen to them.

In spite of all the training and education employees receive accidents still happen. Some do not like the term accidents. A better terminology is system failures. Employees and employers are both subjected to system failures.

The term "human error" is negative sounding and it brings to mind that someone is to blame. When an accident analysis is done the focus needs to be on why the system failure occurred and what part of the system failed, not whose fault it is. Was it a design error, a lack of communication, lack of training, or lack of management involvement? Is the employee overloaded with work either mentally or physically? Did the employee choose to act unsafe because he felt an accident would not happen to him because he has been doing it wrong for twenty years? Once we understand the failure, we can focus on what to do to prevent it from happening again.

Another way to look at human error is to look at the organizational culture. Culture plays a major part in the causes of human error. It is how they were taught or trained to do their jobs, with the older masters teaching the younger workers. Old ways of doing things incorrectly which become habits, repeated and reinforced because that is the way to get the job done, will eventually result in accidents. We are seeing a change taking place in the training of younger workers to integrate safety in their culture. Unions are taking steps to provide more safety training for their members than ever before. Besides the particular craft safely, more members have taken the OSHA 10 hour course than ever before.

It is the culture of the organization to determine what will work within their organization. A company's commitment to safety starts at the top. What keeps it going is the show of continual support from the top and through the ranks.

What determines culture? Here are a few thoughts:

- Who is involved in making decisions regarding safety and safety equipment?
- How much money can or will the company spends to keep safety updated?
- What type of communication is there from the supervisors to the employees on the floor?
- Is teamwork a part of the daily work or is it an "us vs. them" attitude?

By actively being involved with the employees on a day-to-day basis an employer/supervisor/safety personnel shows the employees their commitment to them and to safety. This involvement means talking to them, educating each one on what is safe and what may not be safe and what is definitely not safe.

It is human nature to want someone to care about you. When management shows they care by providing safety systems, tools and equipment, education and an ongoing dialog, the employee will respond in kind by working safely. With this type of work relationship the employee feels no trepidation going to his boss with questions. They know that no question is too stupid.

Often organizations leap at panaceas, quick fixes, which only offer a short lived change. In order to change the safety culture environment and to minimize the human error aspect, leaders need to define the areas they are lacking good safety practices. They need to look at the organizational structure and what may be the best way for their business to improve and change the cultural deficiencies. There are no quick and easy answers to making changes that will benefit the total organization. With change comes uncertainty and there is no refuge from change. Will it work and for how long it will work will plague the decision making process. The bottom line is helping employees through the new changes and seeing the result of good work practices that will result in lower injuries.

SAFETY CULTURE

Today's emphasis is on the safety culture. Safety culture is not primarily about safety, it is about "how we do things in general". It is the tradition in which we are all taught our jobs, our routines. Most workers cannot tell you why they do things the way they do. They just know they have been doing it like this for years and no one has gotten hurt.

This brings up the story about a woman cooking a brisket of beef. She cuts the ends off before putting it in the pan. Her daughter asks her why she did that. Her reply was because she watched her mother do it that way. She called her mother and asked her why she cut the ends off the brisket. Her mother's reply was because her mother did it that way. So they called Grandma and asked her why she cut the ends off the brisket. Her reply was the pan wasn't big enough.

Sometimes we do things because that was the way we were shown. It doesn't make it right.

When safety culture was beginning to become noticed as a new approach to safety, management implemented changes but the safety professionals changed their approach very little. They maintained the same training, same paperwork, the same mindset.

In order to change the safety culture you need to work with people, not make more rules regarding safety. Employees must be involved in the safety process for it to work. The need to keep communications open, get ideas from the employees, and most importantly, listen to what they are really saying will go a long way in understanding what they are thinking and how you can help them change themselves in regards to safety perceptions. On a construction site, ask at the end of any meeting "Are there any concerns or problems?" If you have actively shown an interest in the safety and well being of your employees, you will get responses to this question. It may turn out to be a bitch session, but all you have to do is listen. You can get a lot of information that would otherwise not have been said.

There are four essentials to a safety culture. Act to prevent injuries, Coach one another to identify barriers to safe work practices, Think in ways to support safe behavior and See the hazards. This type of working behavior is called ACTS and it provides knowledge, skills, and tools to fully address the human dynamics of workplace safety.

BEHAVIOR BASED SAFETY IN CONSTRUCTION

The new buzz word/action for safety has become behavior based safety or observed safety behavior. There is a lot of merit with this as it is being proven to be effective for accident prevention. The premise of observed safety is to make both the observer and observed aware of their actions and therefore the consequences of their behaviors.

What we train the observer to look at is critical behaviors. The following is an inventory list of behaviors to observe.

CRITICAL BEHAVIORS INVENTORY

Body Position
Line of Fire
Pinch Points
Eyes on Path
Eyes on Task/Hands
Ascending/Descending

Body Use
Lifting/Lowering
Twisting
Pushing/Pulling
Overextended/Cramped
Response to Ergonomic Risk

Tools and Equipment
Proper Selection/Condition of Tools/Used Properly
Vehicle Selection/Condition of Vehicle/Used Properly
Barricades and Warnings

Procedures
Lockout Tagout Procedures
Communication of Hazard

Pre/Post Job Inspection
Job Assistance

PPE

Head
Eyes/Face
Hearing
Respiratory
Hand
Body Protection
Fall
Foot

Environment

Walking/Working Surfaces
Housekeeping
Storage
Industrial Hygiene

Some doubt that this would be applicable in the construction field. It has been done and it works. One of the key points to making it work is to keep the observation within the individual cultures, i.e, ironworkers only observing ironworkers, bricklayers observing other bricklayers, etc. The different trades on projects represent different cultures. It would not be practicable to ask an ironworker to observe a bricklayer. Here is where culture is a part of the way each craft works. Safety Observation works best within each of the crafts. They know their own ways of doing things so their observations will be geared to their craft.

Observed Safety Behaviors works in general industry as well as in the construction field although there are some limitations, time and money being the biggest factors in the construction arena.

The time factor for training to observe does not come at a small price. Over a period of time the investment in training pays for itself. The knowledge gained and relayed to other employees puts this process at the forefront in ultimate safety training. Employees get satisfaction being a part of the safety process and like it when their opinions are heard and listened to. There is a consciousness of observed behavior, both good and bad, that everyone becomes aware of. This awareness leads to safe behavior

CONCLUSION

OSHA has cited four actions that they feel are the most important for improving workplace safety.
- Worksite analysis
- Hazard prevention and control
- Safety training
- Management leadership.
- Management leadership was deemed the most important.

We cannot let our guard down when it comes to safety. We cannot fully eliminate human error but we can reduce its effects by continual training and reinforcing good behavior. Encouragement is priceless. From top management down to the journeymen, everyone must take an active interest and participate in making the workplace safe. Commitment to words spoken, not just giving lip service, of providing safe working conditions and training is the best investment a company can give itself. Commitment to safety starts at the top and everyone must be involved. Take time for safety. It does make a difference.

ELIMINATION OF HEAT STRESS IN THE GLASS MANUFACTURING ENVIRONMENT

Pat Pride
PPG Industries, Inc.

BACKGROUND AND OBJECTIVE

- Problem: A PPG float glass plant experienced 3 heat stress cases in one year (7 cases total for this business unit)
- Objective: Eliminate heat stress - focus on furnace Hot Repair. Use a technology aimed at *prevention*.
- Assessment Tools:
  - Six Sigma statistical evaluation; catalyst assigned.
  - QUESTemp III Monitor (heart rate & body temp)
  - Polar Monitor (heart rate)

METHODOLOGY

- Develop a Working Group:
  - Plant healthcare representatives: Facility nurse, facility doctor
  - Production representatives
  - Persons to collect, maintain, and evaluate data
- Identify facility "hot work" areas: Areas, jobs, and employees where potential for heat stress exists.
- Initially focused on Furnace hot work areas.
- 15 hot-work employees involved in the assessment; 57 samples.
- Develop a baseline personal heart rate (HR) profile for each hot-work employee using QUESTempIII monitor:
  - All hot repair and back-up hot repair personnel.
  - Others that may be exposed to high-heat jobs.
- Use profile to establish each individual's normal HR baseline.
- Collect HR data trend during heat up/cool down cycle periods while performing hot-work job duties.
- Identify individuals with health risk.
- Investigate cause of health risk before allowing the person to perform high-risk activities.
- Work with facility doctor to determine what is required to allow person to resume hot work activities.
- Institute special requirements, if necessary (i.e., required to wear monitor).

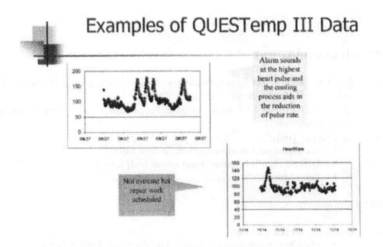

Examples of QUESTemp III Data

RESULTS AND CONCLUSIONS

- Heart rate (HR) indicated reaction to heat much sooner than body temperature response (leading indicator).
- Employees routinely exposed to heat become acclimated to it.
  - This assessment indicated that Hot Repairmen can work longer than other furnace employees. Cool down and recovery are faster.
- While HR monitors were in use, no one experienced heat stress, therefore no data is available during a heat-stress occurrence.
- Cool-Down between work cycles to resume employee's individual normal baseline HR averts heat stress.
- Availability and use of a Cool Room facilitated recovery time to normal HR (allows resuming duties sooner).
- Rule of thumb: 15 minutes IN followed by 15 minutes OUT for Hot Repairmen. *Important* to note this is site, job and employee specific.
- Guideline for maximum allowable HR: 220 minus age (or as determined by facility doctor)
- Other contributing factors to HR response to hot work conditions:
  - Individual variation – 2 employees experienced unexplained rapid HR.
  - General health, medications, etc.
  - Determine individually if fit for duty
- Once baseline is established with QUESTempIII, a HR monitor such as that marketed for exercise (i.e., Polar), is adequate for ongoing monitoring during hot work activity.
- No new heat stress cases since implementing this program (approximately 19 months).

IMPLEMENTATION

- Purchase each Hot Repairman a personal HR monitor (using Polar brand). Nurse sets alarm level for each (220 minus age).
- Hot repairmen pace themselves through use of personal baseline HR and ongoing use of monitor during hot work.
- Cool vests (gel type), various beverages (Gatorade, Propel, water etc.), are also used in conjunction with HR knowledge to prevent heat stress.
- Comments on Cool Room:
  - Convenience & proximity to work location encourage its use.
  - Employees don't have to walk to a remote break room or other area far from the work site.
  - Equipped with conveniences: lighting, benches, marker board, phone, clock, etc.

Cool Room

HEAT STRESS PREVENTIVE PROGRAM SUMMARY

- Establish HR baseline for applicable employees (QUESTemplll)
- Screen for at-risk employees
- Use HR monitor during hot-work activities (Polar)
- Use work/rest cycle (15 min ea or as indicated by HR response)
- Use cool room & A/C areas, cool vests
- Provide suitable beverages & specific diet meals
- Use "observers" of worker behavior; blood pressure.
- Annual training – symptom recognition, prevention, procedures, etc.
- Procedures requiring use of HR monitors

- If anyone experiences heat stress- requires plant doctor to approve return to hot work.
- Plan jobs during cooler periods of day/year if possible
- Have extra manpower for long jobs

DISCLAIMER

Guidelines and protocols described in this presentation have been established based on conditions specific to one facility and its specific operations. It is recommended that other facilities not rely solely on these results but should develop data and programs applicable to their unique operations, work conditions, personnel, and climates.

THE GRAVITY OF GRAVITY – SAFETY'S NUMBER ONE ENEMY

Terry Berg
CertainTeed, a division of Saint Gobain Corporation

This is a brief summary of a multi-media presentation, following which; CDs were distributed at the 67th Glass Problems Conference at Ohio State University on the influences of gravity and the unseen or often overlooked dangers of dealing with it. The complete presentation may also be viewed on the GMIC website (www.gmic.org).

Gravity influences every area of our lives. When you stop to think about it, almost all of the sports we enjoy are nothing more than attempts to defy it or its effects. In baseball, football, or golf we work against it to try to hit, throw or kick a ball farther. In sports like skiing or bobsledding we try to work with it to go faster and faster. The younger generation with their Xtreme sports seems to stick out their tongues at gravity as they loop-the-loop on their skateboards, bikes and motorcycles. Unfortunately, not all landings are happy ones.

I want to touch just briefly upon a few simple aspects of this subject, first from the standpoint of gravity working on us and the objects around us and second from our attempts to work against it.

Falls are the #1 cause of injuries in the home and the workplace and the seriousness of the injury is not necessarily related to the distance fallen, for example, a jet fighter pilot bailed out at 26,000 feet and sustained only a broken hip and a few bruises when his chute didn't open. Below is a chart of industrial accidents within Saint Gobain Corporation over the last two years to further illustrate this point.

- During a demolition of a furnace.
  - Height of 16m : Nose fracture. A miracle! (Feb. 2005)
- In a construction area.
  - Height of 5m : Miraculously, only some contusions ! (Nov. 2004)
- From a scaffolding
  - Height of 4m : Backbone injuries. (March 2005)
- From a platform
  - Height of 4m : Multiples fractures. (March 2005)
- From a product display (sales department).
  - Height of 2m : Wrist fracture. (March 2005)
- From a ladder (x 2).
  - Height of 2m : Foot fracture (April 2005)
    - Lumbar area wound (May 2005)
- From a roll bay storage
  - Height of 2m! : A fatal fall (January 2006)

One of the corollaries to Murphy's Law is that "Nature always sides with the hidden flaw." There's nothing like a fatality to point out the flaws in any system. A direct result of this fatality was the publishing not only of a "Working at Height" control program but also the release of a mandate within CertainTeed that when working at a height of 4 feet or more above the floor, employees must be tied off in accordance with OSHA regulations. Of course, it must be remembered that the OSHA regulations are only designed to prevent impact with the ground so please consider the resultant pendulum effect when placing restraint tie-offs so that swinging bodies aren't flung into other equipment.

The use of proper PPE can only take us so far though. What about falls at ground level when we are not tied oft? Do people in your location still leave cords and hoses on the floor for someone to trip on? Do spills of liquids or other materials sit for hours or sometimes days because of pervasive attitudes like "it's not my job" or "it's no big deal"? Any slip or fall can be deadly under the right conditions. Consider this example: In a non-U.S. facility, three men were guiding an indoor, overhead crane as it was setting down a very heavy load. With the load settled, the operator started to move the crane to pick up another load when one of the cables snagged the corner of the load already in place. This lifted the comer and caused it to start to pivot and tip over. The three men ran to try to stop it from falling over but it was too big. **As** it fell, the top of the frame hit one of the men about waist high and broke his leg and his hipbone as it pinned him to the floor. The force that knocked him down caused him to hit his head and he died in the hospital from a cerebral hemorrhage a few hours later, a fatal fall from ground level!

This brings us to the next danger of gravity, not falling but falling objects. One of our pneumatic batch systems had a badly cast flange burst under load and fall 70 feet right through the roof of the unloading shed nearly missing the truck and making a big dent in the concrete. This could have been a fatality for sure. All of our pneumatic systems now have chain or cable tethers to prevent future like occurrences. It is highly recommended that you evaluate your systems to see if similar preventive are warranted.

There are many reasons why objects fall such as the failure of the support system under excessive load. This was recently witnessed in one plant when in the filling of a new storage bin; the brackets attached to the sides of the bin failed allowing the bin to fall **40** feet smashing everything below. Fortunately, there was no one working inside the building at the time. Objects fall even faster when the support is moved, like the truck that pulled out as the forklift was driving into the trailer. Fortunately the driver of the forklift was wearing his seatbelt and avoided serious injury.

Shifts in the center of gravity may be the most frequent cause of falling objects but I do not have the statistics to back this up. Consider the forklift driver who tries to make a turn in the warehouse with the forks half-way up only to end up tipping over due to the weight and height of the extended mast. It happens! Something else that happens especially in colder weather is dump trucks and even tanker trucks tipping over while they are unloading. The load will freeze to one side of the truck and when the material drains out the other side, the center of gravity shifts and trucks flips and it flips fast. If it is a closed truck, you will never see the problem developing. It was this very problem which killed a man next to a tanker truck in March 2006. Are your unloading areas roped off when trucks are dumping? Will it take a fatality at one of your plants before you get the message'?OSHA hasn't caught on to this danger yet. Have you?

All of us are faced with using outside contractors from time to time and as the following statistics show these are clearly times for heightened awareness.

CONSTRUCTION SITE ACCIDENT FACTS [1]

- On average, there are 1 1,200 disabling accidental injuries every hour.

- More fatalities occur from falls than any other construction activity

- In 1995, there were 6,500,000 construction workers in the United States. Among these workers were 1,040 deaths and 350,000 disabling injuries.

One of the most dangerous pieces of equipment on any construction is a crane. OSHA has recognized this fact and has established very. specific regulations concerning their set-up and use, yet to this day more than 100 fatalities occur every year due to cranes collapsing or tipping over due to shifts in their center of gravity. One story which bears repeating is the story of "Big Blue". This monster crane was being used to lift the sections of the domed roof on to the new Miller Stadium in Milwaukee. Each section weighed more than 200 tons. Cranes. as you well know, are very stable hut only along their vertical, longitudinal axis. Big Blue was no exception with a horizontal. wind thrust rating of only 15 knots. The first lift of the day was made in the morning with three men suspended in a man-basket by a second crane who then successfully anchored the section into place once Big Blue had set it down just as they had done for the previous nine sections. In the afternoon. the wind came up to 26 knots and the safety engineer told the contractor that they had to stop for the day. The construction management said that the work had to proceed. The safety engineer refused, quit and walked off the job. The chief of the three operators running Big Blue also refused and walked off the job just as the safety engineer had done. The second operator told management that he would make the lift. When Big Blue collapsed due to the wind, it fell on the crane holding the three iron-workers and they fell to their deaths. The project was delayed by one year and the punitive damages were $33 million to each of the three families whose fathers were killed. Would you have had the courage to stop the job? We can't fight gravity, but we all have to be willing to fight stupidity!

In conclusion, we must look out for our people and each other. To this end, the GMIC has agreed to host a separate SAFETY section on their website so that we can communicate any newly discovered hazards and recommend safety procedures throughout our industry as quickly as possible. Never let it be said that your company or ours tried to use safety as a means to a competitive edge. Please join us in developing this site to be as useful and timely as possible. I wish to especially thank Pat Pride of PPG and Lauren Alterman of SGC for helping to drive this initiative forward.

Training + Protection + Inspection + Communication + Correction + Compliance = Safety

REFERENCES

1. Accident facts compiled from the Bureau of Labor Statistics, U. S. Department of Labor and the book *Accident Facts*, published by the National Safety Council (1996)

THE LEGACY OF GLASS RESEARCH ACTIVITIES BY THE U.S. DEPARTMENT OF ENERGY'S INDUSTRIAL TECHNOLOGIES PROGRAM

Elliott P. Levine
U.S. Department of Energy, Washington, DC

Keith Jamison
Energetics Incorporated, Columbia, MD

This paper discusses the highlights and accomplishments of the glass-focused energy efficiency activities conducted by the U.S. Department of Energy's (DOE's) Industrial Technologies Program (ITP) since the mid 1990s. These activities include cost-shared research, as well as technology planning, partnership development, and analytical studies. The paper will also explore continuing opportunities for improving energy efficiency in glassmaking operations, the role of the government in promoting collaborative public-private research, and identifying common technology needs among the various sectors of the U.S. glass industry.

INTRODUCTION

ITP is the sole DOE program focused on reducing industrial energy demand. ITP has been engaged in reducing industrial energy consumption through the development of advanced technologies since the late 1970s. Of particular interest to the glass industry was work sponsored by ITP (then known as the Office of Industrial Programs) in oxy-fuel combustion during the 1980s, which helped the eventual commercial development of this technology which is now used extensively throughout the glass industry.

> **ITP Mission and Vision**
>
> ITP's mission is to enhance national energy security, competitiveness and environmental quality by transforming the way U.S. industry uses energy. The ITP vision is an energy efficient American industry that is the global leader in high impact, clean, efficient, and flexible energy technologies and practices.

In the mid-1990s. ITP began a planning effort known as Industries of the Future (IOF). The glass industry was one of nine industries selected to participate in IOF efforts, due to the glass industry's high energy intensity and significant energy consumption – the U.S. glass industry consumes around 250 trillion Btu of energy annually. Significant efforts were conducted in planning and analysis to ensure that Federal investments provided significant energy savings and met technical priorities for the glass industry. These efforts also built on research funded by ITP prior to the IOF effort on oxy-fuel firing, oxygen-enriched air staging and an advanced temperature measurement system, which became commercial technologies.

Since the inception of the Glass IOF program in the mid 1990s, ITP has funded about 35 glass research projects, resulting in both commercial and emerging technologies; two R&D 100 awards (for Oxygen-Enriched Air Staging and Coupled Combustion Space/Glass Bath Furnace Simulation), and significant intellectual property in the form of patents and copyrights. The program has helped the glass industry supplement their limited research funding for process technology, especially in light of the reduction in research staffing and funding at many glass companies which has occurred over the past decade. Also, by sharing the risks and costs associated with developing new process technology, ITP allows technologies to be investigated that otherwise may not be funded.

PARTNERSHIP DEVELOPMENT

From the beginning of the IOF partnership efforts, ITP realized that there was no single organization that represented the technology needs for the glass industry. To maximize the potential for collaboration, ITP encouraged the glass industry to form an umbrella organization to represent its interests. As a result, the Glass Manufacturing Industry Council (GMIC) was formed in 1998. From the original seven founding members, GMIC has grown into a vibrant organization representing around 40 glass companies and other organizations.

To formalize the partnership between GMIC and ITP, a compact was signed by GMIC and ITP in February 1999. This voluntary agreement expressed the intentions of the parties to pursue collaborative efforts through the identification and performance of research, development, and technology demonstrations of mutual benefit to U.S. glass companies and the Nation. The GMIC also served as an advisor during ITP solicitations – providing technology developers a connection with potential industrial partners.

Additionally, an Allied Partnership agreement was signed between DOE and the GMIC in June 2003. This agreement formalized a relationship that will lead to the implementation of numerous energy improvement technologies and techniques across the glass industry. As part of these efforts, GMIC hosted or arranged several energy efficiency training workshops in subjects such as process heating and compressed air.

The GMIC remains a primary conduit for ITP in providing technology solutions to the U.S. glass industry.

Furthermore, ITP is also helping to develop a distance learning class on energy efficiency for the glass industry that will share knowledge to plant personnel. A stakeholder engagement forum, located at http://www.govforums.org/glass/index2.cfm, was also developed to promote dialogue and partnership opportunities on technology issues and solutions, even though its use thus far has been limited.

MAJOR RESEARCH PROJECTS

In the period between 1996 and the present about 35 major research projects were conducted. These projects were focused primarily on improving the energy efficiency of furnace operations, which is common to virtually all glass sectors and usually accounts for the largest proportion of energy consumption. Efforts included furnace modeling, refractory corrosion, and alternative melting techniques – technologies that could be applied to multiple glass sectors. Additional research and development included: efforts on sensors and controls, extending oxy-fuel technology, and preheating of batch and cullet.

The glass industry cost-sharing on these projects amounted to around 35% of total project costs. The vast majority of these projects were selected through competitive solicitations, and annual review meetings were conducted to ensure that projects were on track to meet their technical objectives. Brief summaries of recent and successful projects are depicted in the table below by technical area, and include both the activities undertaken and their potential significance (with project completion year in parentheses). Other projects that were funded by the program are shown in Appendix A. And fact sheets or final reports on these projects can be found at http://www.eere.energy.gov/industry/glass/portfolio.html.

Next Generation Melting Systems		
Next Generation Melting	• **Energy-Efficient Glass Melting: The Next Generation Melter:** Researchers have designed and are fabricating a pilot scale submerged combustion glass melter. Two series of tests will be performed, and product glass properties will be analyzed. (ongoing)	• This novel melter design can significantly reduce furnace size and capital costs along with high thermal efficiency and low gas-phase emissions.
	• **High-Intensity Plasma Glass Melter:** Investigators have developed a plasma-melting system that is generically suited to melting a large variety of glass compositions. A prototype system has been constructed and tested, and product quality was evaluated. (2006)	• This modular melter can provide a low systems cost for selected glass applications and rapid product changes.
Energy Efficiency Performance Improvements		
Furnace Modeling	• **Development and Validation of a Coupled Combustion Space/Glass Bath Furnace Simulation:** Researchers have developed a validated glass melting furnace simulation model that incorporates innovative features. The combustion space and glass bath models are coupled at their interface through the use of appropriate heat flux and temperature continuity conditions. The combustion space model incorporates a rigorous treatment of the radiative heat transfer to the glass bath. (2006)	• A modeling solution that can better regulate heat flux distribution on the batch and glass melt surfaces both in existing and new glass furnaces.
	• **Development of an Energy Assessment Protocol:** The objective of this project was to monitor and characterize furnace operations, identify potential energy inefficiencies, recommend energy saving changes, and implement changes and re-evaluate energy consumption; in order to optimize oxy-fuel firing in glass furnaces. (2003)	• A best practice for identifying near-term energy saving opportunities in oxy-fuel furnaces.
	• **Modeling of Glass Making Processes:** This project developed data for modeling of glassmaking processes. The properties of interest included thermal conductivity, thermal diffusivity, specific heat, density, surface tension, viscosity, electric conductivity, gas solubility, diffusion coefficients, and elastic constants all in the temperature range of 700° C to 1600° C. (2003)	• A robust, published database available to the glass industry that can be used to improve modeling capabilities.
Materials-Related	• **Diagnostics & Modeling of High-Temperature Corrosion of Superstructure Refractories in Oxy-Fuel Glass Furnaces:** Researchers worked to identify factors controlling refractory corrosion in oxy-fuel glass furnaces, developed models to predict corrosion rates based on these factors, and developed in-situ optical techniques to monitor gas-phase alkali concentrations in glass furnaces. (2003)	• Knowledge that can be used to design furnaces and refractories with lower corrosion rates.
	• **Improved Refractories for Glass:** This project developed key data needed for advanced refractories. Laboratory testing, analysis, and characterization of a variety of refractory materials was conducted in order to develop refractories with superior corrosion and creep resistance in melting tanks in all segments of the glass industry. (2000)	• Knowledge that was used in the development of refractories with advanced physical properties.
Combustion Systems	• **Development/Demonstration of an Advanced Oxy-Fuel Fired Front-End System:** Scientists have developed a novel, oxy-fuel fired front-end system. After laboratory design testing of burners and system control, the system was demonstrated in a commercial fiberglass production plant. (2006)	• Oxy-fuel technology extension into the front-end, with fuel savings greater than 50%.

	• **High-Heat Transfer Low-NOx Natural Gas Combustion:** Researchers developed and demonstrated a high luminosity, low NOx burner that internally modifies the fuel prior to combustion, as well as controls fuel/air mixing to substantially increase the formation of soot within the flame. This process enhances the luminosity of the flame, increasing the heat transfer rates and decreasing the flame temperatures. (2003)	• Flat flame burner available for sale that was designed for oxy-fuel furnaces and provides increased thermal efficiency and low NOx emissions.
Non-Furnace Operations	• **Process Optimization Strategies, Models, and Chemical Databases for On-Line Coating of Float Glass:** This project developed modifications to atmospheric pressure chemical vapor deposition to increase the efficiency of reactant utilization. Researchers conducted detailed studies of the underlying deposition process. Computational models were developed that can predict defects in coatings, and pilot-scale testing was conducted. (2005)	• A model and kinetic data developed to improve coating efficiency and reduce coating material use.
Sensors and Process Control	• **Measurement and Control of Glass Feedstocks:** Researchers developed a probe based on laser-induced breakdown spectroscopy to measure the chemical makeup of glass feedstocks in real-time. Artificial neural network software provides high-speed analysis of data. (2006)	• An instrument designed to quickly detect contaminants and batch nonuniformity in raw materials and cullet.
	• **Improvement of Performance and Yield of Continuous Glass Fiber Drawing Technology:** This project applied six sigma quality methodology and fundamental glass science to reduce fiber breakage and resultant waste. During the project, instrumentation was developed, simulation models were employed, and the process was optimized. (2006)	• Reduce break frequency for fiber drawing by a factor of four, leading to less downtime and increased throughput.
Environmental Performance	• **Monitoring and Control of Alkali Volatilization and Batch Carryover for Minimization of Particulates and Crown Corrosion:** Researchers collected data to determine the conditions having the greatest influence on volatilization, batch carryover, combustion efficiency, and furnace efficiency. A prototype measurement instrument using laser-induced breakdown spectroscopy has been designed and built. (2005)	• A sampling probe which provides data that can be used to assist in reducing crown corrosion and particulate emissions.

GLASS PROJECT LABORATORY USER SERVICES (GPLUS)

The GPLUS program, which started in 2000, enabled many glass manufacturers to conduct small-scale scoping studies in collaboration with any of the DOE national laboratories – facilitating access to the vast technical capabilities and technologists found in the DOE national laboratories. The program was cost-shared by ITP and the glass manufacturer, and results have been provided to the GMIC to encourage technology transfer.

Over the past several years, nearly 30 of these scoping studies – roughly equivalent to one month's time in the laboratory – were conducted in a broad range of subject areas related to improving energy and production efficiency, including glass properties, process behavior, sensors and process control, and melting and modeling. A listing of these studies is shown in Appendix B.

PARTNERS AND PARTICIPANTS

Through its interactions with the glass industry, ITP has been able to attract major industry players to participate in research and planning activities. Approximately 100 participants throughout the Nation have partnered with ITP over the past decade. Participation ranged from conducting research and providing specialized technical knowledge to providing cost-sharing assistance and facilities for demonstration testing. Participants include large glass manufacturers, niche glass manufacturers, industry vendors and technology suppliers, academic institutions, national laboratories, and other partners. A listing of known partners and participants is shown in Appendix C.

TECHNOLOGY PLANNING

Technology planning activities identified the highest priority needs in the glass industry and guided R&D efforts conducted by ITP, and included the development of the glass industry vision and technology roadmap – which ITP helped facilitate and the glass industry provided input. The glass vision document, "Glass: A Clear Vision for a Bright Future," was published in January 1996. This document spelled out primary goals and priorities, including a goal of reducing the gap between actual energy use and the theoretical minimum by 50% by the year 2020. A technology roadmap workshop was conducted in 1997, which ultimately led to the publishing of the "Glass Industry Technology Roadmap" in April 2002 by the GMIC. The roadmap identified technical barriers and priority research needs in four technical areas: production efficiency, energy efficiency, environmental performance, and innovative uses.

Vision Goals (2020)

- Operate with production costs at least 20% below 1995 levels;
- Recycle 100% of all glass products in the manufacturing process,
- Reduce process energy use from present facility levels by 50% toward theoretical energy use limits;
- Reduce air/water emissions by a minimum of 20% through environmentally sound practices;
- Recover, recycle and minimize 100% of post consumer glass where consumption is greater than 5 lb/capita;
- Achieve six sigma quality through automation, process control, optimized glass composition/strength, and computer simulation;
- Create innovative products that broaden the marketplace;
- Increase supplier and customer partnerships in raw materials, equipment, and energy improvements.

Source: http://www.eere.energy.gov/industry/glass/pdfs/glass_vision.pdf

Source: http://www.eere.energy.gov/industry/glass/pdfs/glass2002roadmap.pdf

Other efforts in technology planning included several technology workshops. Sandia National Laboratories led an effort to produce a roadmap for glass coatings. Two workshops were conducted that focused on oxy-fuel firing. and another workshop focused on next generation melting – which served to define the level of industrial interest in collaborating in this area.

ANALYSIS

Analytical studies were conducted by ITP to help identify opportunity areas and provide baseline data. These studies helped overcome the lack of benchmarking data in the glass industry. The Energy and Environmental Profile of the U.S. Glass Industry provides an analysis of energy consumption by process. and an overview of various processes used in *the* glass industry "Glass Melting Technology: A Technical and Economic Assessment" provides insights on previous efforts for advanced melting technologies conducted both in the United States as well as many that have been investigated overseas. And the "Industrial Glass Bandwidth Analysis" provides an assessment of potential energy savings from implementing state-of-the-art technology as well as potential technology improvements, as depicted in the accompanying figure. A second-stage bandwidth effort will take a closer look at the potential impacts of individual technology improvements and applications. These publications not only provide insight into glass manufacturing technology. but provide a sound basis for the pursuit of individual research topic areas

Sources: http://www.eere.energy.gov/industry/glass/pdfs/glass2002profile.pdf
http://www.gmic.org/News/ISBN0976128306txt.pdf

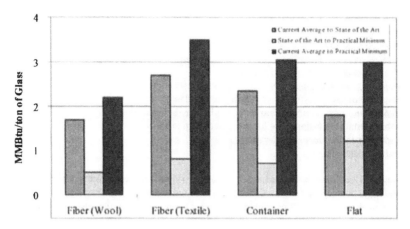

BANDWIDTH SUMMARY
Total Potential Energy Reduction per Ton Produced For Glass Melting/Refining When
Converting From (a) Current to State of the Art, (b) State of the Art to Practical Minimum. and
(c) Current to Practical Minimum

FUTURE OPPORTUNITIES

While the future of the ITP Glass program is unclear at this time, ITP will continue to provide
many opportunities for improving energy efficiency in the glass industry. In crosscutting
technology, other current ITP technology areas such as materials and sensors offer opportunities
for technology development – current or recent projects include thermoelectric materials for
electricity generation from furnace waste heat and flame image analysis for furnaces. In the
coming years, efforts may focus more on expanding fuel flexibility and recovering waste heat
throughout industry.

In addition, there are several opportunities for adopting Best Practices – for example, several
glass manufacturers have recently participated in ITP-sponsored energy saving plant
assessments, and many energy-saving software tools are available from ITP. Another round of
energy saving plant assessments will be offered in late 2006. The robust ITP website
(http://www.eere.energy.gov/industry) will be a valuable source of information about these
technical and financial opportunities and available technology solutions. To further industrial
outreach, a workshop on energy reduction in the glass industry is being planned for early 2007.

CONCLUSIONS

Over the past decade, ITP has successfully partnered with the glass industry. Overall, the ITP
program has made a significant difference to the glass industry by:
• *Providing new knowledge* such as the high-temperature glass properties database.
• *Developing new technology* including four ITP-sponsored commercial technologies: oxy-fuel
   firing, oxygen-enriched air staging, advanced temperature measurement system, and high-

luminosity low-NOx burner. Several other technologies have the potential to emerge into the marketplace.

- *Solving technical barriers and problems* through the research and development of technology.
- *Providing energy savings and energy saving protocols* though the application of commercial technologies and protocols in the marketplace.
- *Developing new learning tools* such as a distance learning class under development to share knowledge of energy efficiency opportunities to plant personnel in the glass industry.
- *Encouraging new collaborations* by providing the requisite framework and helping the GMIC establish itself as a major agent for broad collaboration within the glass industry.
- *Identifying future opportunities* through the development of the technology roadmap and bandwidth documents.

However, ITP realizes that technical success alone will not ensure market acceptance of technologies developed. Increased efforts are required to ensure that adequate business planning is conducted, the value proposition is appropriate for end-users, and that there are champions for the technology. Otherwise, the value of the technology will not be maximized.

Finally, to ensure the legacy of the ITP glass program, a dedicated website is under development to archive research findings, program documents, and other results. The URL for this website will be http://www.osti.gov/glass. This website will be a valuable repository of knowledge for the entire glass community.

APPENDIX A: OTHER ITP RESEARCH PROJECTS

The table below depicts other ITP research projects that were not discussed previously. Some of these projects were technically successful, but are unlikely to move further forward into the marketplace without additional development efforts.

Energy Efficiency Performance Improvements	
Furnace Modeling	• **Glass Furnace Combustion and Melting User Research Facility:** This project designed a state-of-the-art user facility for researchers. The objective was to provide diagnostic tools, measurements, and analysis of the flows and chemical reactions, for purposes of improving combustion control, product uniformity, and refractory life. However, the facility was never constructed. (2002) • **Development, Experimental Validation, and Application of Advanced Space Models:** Researchers developed combustion space models for both air-fuel and oxy-fuel furnaces. Experimental data was generated and validated in float furnace operations, and the model was used to improve the efficiency of a float furnace during the rebuild process. (2000)
Materials-Related	• **Molybdenum Disilicide Composites for Glass Processing Sensors:** Scientists evaluated the production of molybdenum disilicide hybrid composite tubes and coatings for thermocouple sheath applications. Advantages of molybdenum disilicide tubes over conventional materials include: electrically conductive, stronger than ceramic refractory materials, and chromium-free. Plasma-spray forming techniques were established and optimized for creating these tubes. (2001) • **On-Line Chemical Vapor Deposition for Composites:** Researchers used a research reactor to determine identities and amounts of gaseous phase species present during chemical vapor deposition (CVD). The purpose was to investigate the CVD process used to coat float glass to determine the source of some present difficulties, to optimize the process, and to develop new coatings. (1999) • **Synthesis and Design of Intermetallic Materials:** Researchers developed processing methods for molybdenum disilicide ($MoSi_2$) based materials, such as plasma spraying/spray forming and electrophoretic deposition. Activities also included development and characterization of $MoSi_2$-based high temperature structural materials with focus on $MoSi_2$-based composites, and plasma sprayed $MoSi_2$ materials and microlaminate composites. (1998)
Batch and Cullet Preheating	• **Electrostatic Batch Preheater System:** Researchers developed an advanced technology to preheat mixed glass batch and abate emissions from the waste gases of an oxy-fuel fired furnace. A full-scale demonstration of the technology was planned, but demonstration host site difficulties resulted in early termination of the project. (2002) • **Integrated Batch & Cullet Preheater System:** This project attempted to demonstrate the technical viability, system reliability, and economic benefits of batch and cullet preheating using raining bed preheating technology for both conventional preheated air and oxygen-fuel combustion. Demonstration host site difficulties resulted in early termination of the project. (2000)
Non-Furnace Operations	• **Enhanced Cutting and Finishing of Handglass Using a Carbon Dioxide Laser:** Scientists developed laser-enhanced cutting and finishing methods to dramatically decrease waste and improve productivity in the manufacture of handblown glass. A bench-scale prototype system was constructed using a sensor-controlled, moderate-power, CO_2 laser to precision-cut the glass and produce a finished edge. The portable prototype was demonstrated in industrial facilities. (2002) • **Integrated Ion Exchange Systems for High-Strength Glass Products:** Researchers evaluated integrated ion-exchange systems for soda-lime based composition in which strengthening times could be reduced by at least a factor of 2, and up to a factor of 5. Large potential markets exist in architectural, automotive, and specialty applications, but remain untapped due primarily to the high cost of chemical strengthening. (2001)
Sensors and Process Control	• **Advanced Process Control for Glass Production:** Scientists developed and implemented an advanced process control system that should increase the amount of high-quality product from the initial manufacturing step. The system integrated: (1) a model that relates process parameters (e.g., temperature, deformation, cooling rate) to final product quality; (2) a suite of novel, 3-D stress and temperature sensors for measuring process parameters; (3) a system for integrating and analyzing data from a wide range of sensors; and (4) cognitive control software for adjusting the process parameters to maintain product quality. (2002) • **Auto Glass Process Control:** This project developed a novel non-contact stress measurement method for improving quality control of automotive glass. A patent has been submitted for the method. In addition, a computer code for glass product forming was validated. Pacific Northwest

	National Laboratory is working with a software vendor to include the code in their commercial software program. (2002)
	• **On-Line Sensor System for Monitoring the Cure of Coatings on Glass Optical Fibers and Assemblies:** Scientists worked on developing a continuous, non-contact sensor to monitor coating cure at line speeds of up to one kilometer per minute. This sensor uses infrared spectroscopy to inspect 100% of the polymer coatings to assure proper curing. The information could be used to create a visual display, sound an alarm, or activate a feedback loop for process control. (2000)
	• **Dynamic Expert Systems Control for Optimal Oxy-Fuel Melter Performance:** Researchers attempted to develop an innovative dynamic expert system of controls for oxygen-enriched glass melting plants. Novel, non-intrusive laser-based diagnostic methods for temperature measurement were investigated. (2000)
	• **Redox State Sensor Technology in Glass Melts:** Researchers attempted to develop a robust sensor for reliably measuring the ratios of various redox components in a melt during processing. A redox sensor could then be integrated with a thermochemical model for predicting glass properties based on the chemistry of the melt and a control strategy for adding reductants or oxidants during glass production to improve the final product. (2000)
Environmental Performance	• **Development of a Process for the In-House Recovery and Recycling of Glass from Glass Manufacturing Wastes:** This project evaluated a separation technology for purifying and upgrading glass waste streams. Researchers studied imperfections and tested thermal and chemical methods to purify and upgrade the waste glass. An economic evaluation was conducted to ensure that the most technically efficient and cost effective method is chosen for further process development. (2003)

APPENDIX B: GPLUS PROJECTS
A listing of GPLUS projects is provided in the table below:

Glass Properties	• Application of Strength Test Data to the Design of Automotive Glazing With and Without Holes • Glass Properties Determination • High Temperature Viscosity Measurements • Mechanical Strength Studies of Automotive Tempered and Annealed Glass with Holes • Strength of Tempered Glass with Holes Including Stress Corrosion Cracking • Thermal Conductivity Measurement of Fiberglass • Thermophysical Properties of Glass
Melting and Modeling	• Application of Furnace Model to Oxy-Fuel Furnace for the Production of Amber Glass • Improvement of Oxyfuel Burner Design and Operations • Pollution-Free Fiber Glass with Reduced Melting Temperature
Sensors and Process Control	• Construction and Performance of an Aerosol Generation System for Calibration of In-Situ Laser Induced Breakdown Spectroscopic Measurements • Development of New Control Software for Integration of an Echelle Grating Spectrometer for Use in Laser Induced Breakdown Spectroscopic Measurements • High Temperature Thermocouple Degradation Study • LIBS as a Glass Melt Monitor • Millimeter Wave Diagnostics for Glass Fiber Drawing • Spectral Analysis and Imaging of Colored Glasses
Process Behavior	• Analysis of Refractory Metal Needles Formed During Molten Glass Extrusion • Asymmetrical Windshield Construction • Batch Reactions of a Soda-Lime-Silicate Glass • Dependence of Boron and Sodium in Flue Gas from a Borosilicate Glass Melting Furnace on Oxygen to Natural Gas Ratio • Dependence of Boron and Sodium in Flue Gas from a Borosilicate Glass Melting Furnace on Conditions in the Caustic Spray Tower and Electrostatic Precipitator • Foaming of E-Glass • Glass Coating Enhancement • Reduction in Corrosion of Refractories Using Encapsulated Glass Additives • Sulfate Fining Chemistry in Oxidized and Reduced Soda-Lime-Silica Glasses • Using Colored Cullet for Making Beautiful Glassware

APPENDIX C: PARTNERS AND PARTICIPANTS
Known partners and participants in the Glass IOF program are depicted below:

Glass Companies		
	• AFG Industries	• Osram Sylvania
	• AGY	• Owens Corning
	• Canandaigua Wine	• Owens Illinois
	• Cardinal Glass	• Pilgrim Glass
	• CertainTeed	• Pilkington LOF
	• Corning	• Plasmelt Glass Technologies
	• Davis Lynch Glass	• PPG Industries
	• Emhart Glass	• Saint-Gobain Containers
	• Fenton Art Glass	• Saint-Gobain Vetrotex America
	• Fire & Light Originals	• Saxon Glass Technologies
	• Gallo Glass	• Schott Glass Technologies
	• GE Lighting	• Sierracin
	• Johns Manville	• St. George Crystal
	• Kopp Glass	• Techneglas
	• Leone Industries	• Thomson Consumer Electronics
	• Libbey	• Viracon
	• Longhorn Glass	• Visteon
	• Marble King	• Vitro
Industry and Technology Suppliers	• A.C. Leadbetter and Sons	• Maxon
	• AccuTru International	• McDermott Technologies
	• Advanced Control Systems Inc.	• Merkle Engineers
	• Air Liquide	• Monofrax
	• Air Products and Chemicals	• NARCO/Harbison-Walker (now RHI
	• BOC Gases	Refractories)
	• Corhart	• New York Gas Group
	• Eclipse/Combustion Tec	• Physical Optics Corp.
	• Engelhard	• Plasma Processes
	• Energy Research Co	• Praxair
	• Exotherm	• Thermex
	• Fluent	• Thermo-Power
	• Gas Technology Institute	• TransResources
	• Laidlaw Drew	• U.S. Borax
	• Lilja	
Academia	• Brigham Young University	• Ohio State University
	• Center for Glass Research (Alfred University)	• Penn State University
	• Clark Atlanta University	• Purdue University
	• Cleveland State University	• University of Alabama-Birmingham
	• Georgia Institute of Technology	• University of Missouri-Rolla
	• Mississippi State University	• University of Utah
		• West Virginia University
National Laboratories	• Ames Laboratory	• National Energy Technology Laboratory
	• Argonne National Laboratory	• National Renewable Energy Laboratory
	• Idaho National Laboratory	• Oak Ridge National Laboratory
	• Lawrence Berkeley National Laboratory	• Pacific Northwest National Laboratory
	• Los Alamos National Laboratory	• Sandia National Laboratories
Other Participants	• Energetics	• N Sight Partners
	• Glass Industry Consulting	• New York State Energy Research and
	• Glass Manufacturing Industry Council	Development Authority
	• Henry Technology Solutions	• Society for Glass Science and Practices
	• JFM Consulting	• Shell Glass Consulting
	• Laboratory of Glass Properties	• Warren Wolf Consulting

MODELING

APPLICATION OF RIGOROUS MODEL-BASED PREDICTIVE PROCESS CONTROL IN THE GLASS INDUSTRY

O.S. Verheijen, O.M.G.C. Op den Camp, TNO Glass Group, Eindhoven, The Netherlands
A.C.P.M. Backx, Eindhoven University of Technology, The Netherlands
L. Huisman, IPCOS, The Netherlands

ABSTRACT

The steady state and dynamic behaviour (heat transfer, temperatures, glass and gas flows) in glass furnaces and forehearths can be described accurately and reliably by Computational Fluid Dynamics (CFD) models [1] such as the TNO Glass Tank Model (GTM-X). Application of these detailed, but also slow models for direct on-line control or optimization of glass melting processes is not possible without strong model reduction. TNO, IPCOS and the Eindhoven University of Technology have developed a generic approach, the so-called Proper Orthogonal Decomposition (POD), which is able to reduce the complex CFD glass furnace simulation model to no more than approximately 50 equations, while maintaining the required accuracy and level of detail. Herewith, the computational speed of the reduced order model increases drastically even up to 10.000 times faster than real-time. By following this approach, the resulting reduced models have become so fast, that they can directly be applied in Model based Predictive Control (MPC). This paper describes the benefits of the so-called Rigorous Model based Predictive Control system (rMPC: an MPC based upon a fast, detailed, and accurate 3D CFD model). Also, the approach for setting-up such a controller is discussed and results are shown of an rMPC installed at a container glass furnace to control glass melt temperatures.

INTRODUCTION

Currently, most glass melting furnaces are controlled manually. In this conventional approach to keep a furnace at the desired state, the glass production process is controlled via PID set-points imposed to control variables (CV's) such as primarily temperatures for crown, furnace bottom, and individual feeder sections. In general, PID set-points for given process conditions (pull rate, glass colour, etc) are derived from practical experience from glass furnace operators. As optimal CV set-points depend on process conditions, production of new products requires the (time-consuming) empirical determination of new optimal PID set-points. A second major disadvantage of conventional furnace control is that this type of control is purely feedback: conventional controllers are not capable of feed forward control on disturbances upstream. For example, the effect of temperature fluctuations measured at a feeder entrance on glass melt temperatures downstream are only then compensated (by adjusting PID set-points) when thermocouples downstream (which are controlled by a PID controller) measure these fluctuations. The time delay for anticipation on measured disturbances upstream ranges from several hours up to even more than a day (dependent on positioning of sensors and residence time distribution). Similar to optimal values for PID set-points, product changeover strategies and control actions are based upon operator experience, which generally results in non-reproducible furnace control over shifts and time. To attain reproducible furnace control, automatic control of glass melt temperatures (or any other variable that determines furnace performance) is almost indispensable and can lead to considerable energy savings and improved and more stable glass quality.

In the recent years, process control systems have become available for automatically control of important temperatures, taking over this part of the job of an operator, who can spend his/her time now on other important tasks such as furnace maintenance. Since PID controllers are much to slow and have no predictive capability (i.e. do not give feed forward response on disturbances to keep the temperature within a limited range), model based predictive control (MPC) is being used. With MPC, PID set-points for CV's are based on predictions from a validated (simulation) model of the glass furnace melting tank, feeder or any other part of the furnace. A large advantage over conventional control is that MPC is predictive, which means that MPC is able to anticipate in an early stage on measured process disturbances such as redox and temperature variations. The simulation model predicts and MPC applies optimal PID set-points for disturbance rejection, load and/or colour changeovers, and new products. As with MPC control actions are based on models, these actions are reproducible, whereas with conventional control the control actions are considered as operator dependent. Finally, in contrast to conventional control, with MPC control is based upon multi-objective optimisation: while during conventional control only one single (averaged) temperature is kept stable, with MPC a total crown profile or melter bottom temperature profile can be prescribed (MIMO = multi input multi output).

With respect to MPC two types can be distinguished, viz. 'black box model based predictive control' (bMPC) and 'rigorous model based predictive control' (rMPC). Models for bMPC are experimentally derived dynamic models that are determined by step tests or Pseudo Random (Binary) Noise Sequence (PRBNS/PRNS) tests on the real furnace. In contrast, rMPC models are derived from industrial validated three dimensional (CFD-based) simulation models. Rigorous model based predictive control has some important advantages over black box based model based predictive control:

• Derivation of models for bMPC requires industrial tests. By interference in the actual glass-melting tank and/or feeder, flow pattern and temperature distributions are disturbed, which sometimes offsets the process and may lead to increased project reject and expensive man-hours to re-stabilize the process. With rMPC, control models are derived from mathematical simulation results and do not require interference in the real process.

• Both bMPC and rMPC relate manipulable process inputs (MV's) to measurable and to be controlled process responses (CV's). Additionally, rMPC allows soft-sensor functionality: as three dimensional (CFD-based) simulation models contain complete information on temperature distribution, flow patterns, and distribution of all redox active species throughout the furnace, critical (non-measurable) parameters affecting glass quality and process performance such as hot spot temperature, spring zone position, intensity of the circulation rolls, fining zone volume, and (minimum) residence time (distribution) can be determined. Therefore, with rMPC relations between manipulable process inputs (MV's) and soft-sensor data can be derived and control strategies and optimisation can be more directly related to glass quality and furnace performance.

• As models for bMPC are derived from measured responses on small variations in MV's (using PR(B)NS) around a certain working point, these models will probably be not valid after a significant change in working point. In general, a new working point requires set-up of a new black box model resulting in an additional interference in the production process. In contrast to bMPC, the working envelope for rMPC is large: the response on CV's by large changes in MV's can easily be determined using industrially validated three dimensional (CFD-based) simulation models, which is less time-consuming than laborious industrial derivation of black box models for a large series of working points.

Because of the advantages of rMPC over bMPC, TNO started the development of rigorous Model based Predictive Control for glass-melting furnaces and feeders. However, computations of detailed 3-D simulation models (serving as basis for rMPC) are very time-consuming: one steady state simulation of a complete furnace including refiner(s) typically takes about a day, which indicates that the currently used detailed 3-D simulation models are not suitable for MPC. This was one of the reasons to develop simulation tools that are much faster with a high level of reliability. These fast simulation tools open up a wide variety of applications of glass furnace models. In the framework of a Dutch EET-project, TNO Glass Group and IPCOS in cooperation with the Eindhoven University of Technology have developed a method to set-up fast Glass Process Simulators and based on these simulators, ʀMPC-controllers that enable control for a very large working envelope.

APPROACH TO SET-UP CONTROL MODELS BASED UPON RIGOROUS SIMULATION MODELS

- The approach starts with setting up a detailed 3-D CFD model for the furnace under consideration. This simulation model is of course thoroughly evaluated and validated, as the performance of the controller will depend largely on the quality of the underlying model.

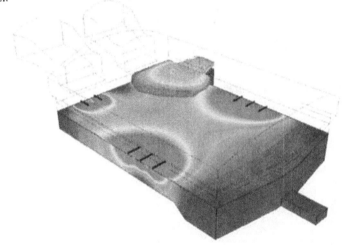

Figure 1: Simulation result, the distribution of the energy release by electric boosting in combination with the shape of the batch blanket from a side charger.

- Subsequently, dynamic tests (PRBNS) are performed upon the simulation model instead of applying (small) step changes on the real furnace. The simulation tests couple the modeled response of temperatures and flows in the feeder to modeled changes in the input. The special mathematical technique Proper Orthogonal Decomposition of temperature field predictions [3], is applied to increase the speed of the modeling tasks and consequently the currently applied reduced models are very fast (i.e. about 10000 times real time). For a detailed description of POD is referred to [4].
- To determine optimal control input parameters (in this case set-points electrical boosting and combustion), enabling the achievement of a constant throat temperatures, the Linear Quadratic Regulator (LQR) type of controller has been applied. Using LQR, minimization of a multiple objective controller cost function is carried out. The controller cost function contains prioritized objectives and constraints. The constraints relate amongst others to minimum and maximum crown temperatures. The main control objective concerns the desired temperature at the melter throat.

APPLICATION OF AN rMPC TO A CONTAINER GLASS FURNACE

The performance of rMPC is demonstrated here by the application to a container glass furnace producing different colours of glass: amber and green. The furnace is a gas fired end-port furnace. Additional heating to the glass melt is provided by an electric boosting system with two transformers. One transformer provides power to support batch melting, the other transformer is used for barrier boosting at the front end of the furnace. The furnace is equipped with two batch chargers from the side at the rear end and has one throat. It provides glass for three production lines. In the bottom of the melter bath, 4 thermocouples have been installed, one of which measures the temperature in the stone close to the bottom of the glass bath at the throat entrance. According to the glass producer, the temperature as measured by this thermocouple shows a high correlation with the glass quality produced by the furnace. A detailed three-dimensional CFD model has been set-up for the furnace, in which the heat exchange between the combustion space and the melter was described on the basis of crown temperature measurements. A reduced model of the furnace was constructed. The following model inputs were considered: the furnace pull, the percentage of cullet added to the batch, the weight percentage of the colouring agents in the glass melt, the power of the two boosting transformers and three crown temperatures. Simulations were performed for different representing working points, results were stored in a snapshot matrix, from which a reduced model was set-up. In half a second, the reduced model computes the three-dimensional temperature distribution in the glass melt and refractories for a real-time period of 1 hour. In half a minute a colour changeover of 3 days duration can be simulated in detail within the required modelling accuracy (see Figure 1).

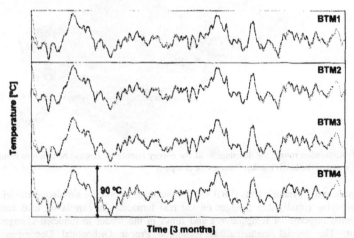

Figure 1: Results of the reduced model (in red) versus the results of the detailed CFD model (in blue) for the temperature in the four bottom thermocouples in the melter over a period of 3 months.

The figure above shows the comparison between the results of the ultra fast reduced model and the original CFD model for the furnace. At the position of the bottom thermocouples, the maximum differences between the two models are in the order of 2 °C on a range of 90 °C, which means that also the reduced model is valid for a very large working range (the working area for the CFD model is in principle not limited as far as the model inputs are physically realistic). In other areas of the mathematical domain, a similar accuracy of the reduced model is achieved.

As off-line tests have shown that the reduced model is applicable for large changes in the model inputs, the model can be used for online control. First, the reduced model has been incorporated in the INCA control environment of IPCOS. The rMPC controller that has been established in this way has been connected with the furnace's Distributed Control System (DCS) by means of an OPC connection.

With the rMPC controller, the following results have been obtained so far:

- Identification of the most-sensitive MV for optimal stabilization of the throat temperature.
- The MV's are more active (still within constraints); the freedom in varying MV's to achieve the most stable throat temperature is maximal.
- Reproducible control as the controller performance is not influenced by operators and shifts.
- All bottom temperatures are stabilized; a reduction in bandwidth of temperature variations at the throat of 75 % has been achieved.
- Although energy and cost optimisation has started only recently, a reduction in energy consumption of 1% has been reported as first result.

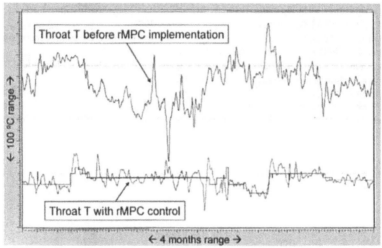

Figure 2: Temperature variations with time for a thermocouple installed in the throat of the furnace, before (blue) and after (red) application of the RMPC controller. For the RMPC controlled furnace also the set point changes are given. The controller is switched on also during colour changeovers.

Figure 3 shows the temperature variations in the throat of the furnace, before (blue curve) and after application of RMPC (red curve). Clearly, the reduction in bandwidth of the temperature variations is shown, although the controller is also used during colour changeovers where larger temperature variations are inevitable.

CONCLUSIONS
Very good results have been achieved with rMPC (rigorous model based predictive control) in which a reduced model based on a detailed CFD model is used to control the temperatures in furnaces and forehearths. Temperature variations in furnaces were reduced to a bandwidth of 8 °C for one colour and to less than 4 °C for another colour, whereas in the uncontrolled situation the bandwidth was more than 20 °C (colour changes excluded). In forehearths the bandwidth was reduced to less than 1 °C (container glass), even in situations that the production process was subject to relatively large disturbances (redox variations in recycling

cullet). Currently the demonstrated rMPC controller is even used during colour changeovers, which shows the confidence of the glass manufacturer in the system and its performance.

The application as shown in this paper is only one out of several examples. It very clearly shows that Proper Orthogonal Decomposition is the missing link between detailed CFD simulations and the required fast models that are required for online applications and more specifically for control. A generic approach to construct ultra fast reduced models has been developed and demonstrated.

The POD technique results in reduced models that give the same level of detail as the original CFD model, but which only need a small fraction of the computational effort. The number of possible applications of such models is unlimited: from direct application in furnace control to online furnace monitoring. Such models are fast enough to be incorporated into process optimization tools or operation support systems; to supply information to furnace engineers or glass technologists on daily furnace operation in a continuous drive to improve yield, to decrease cost, to decrease the consumption of energy, to lower emissions and first of all to stay out of production problems.

REFERENCES

[1] Krause, Dieter and Loch, Horst (Eds.) "Mathematical Simulation in Glass Technology", Schott Series on Glass and Glass Ceramics, Mainz, 2002.

[2] Astrid, P.: "Reduction of process simulation models, a proper orthogonal decomposition approach". PhD thesis, Technische Universiteit Eindhoven, 2004.

[3] Huisman, L.: "Control of glass melting processes based on reduced CFD models". PhD thesis, Technische Universiteit Eindhoven, 2005.

[4] Op den Camp, O.M.G.C., Verheijen, O.S., Backx, A.C.P.M., Huisman, L., 'Application of Proper Orthogonal Decomposition to reduce detailed CFD models of glass furnaces and forehearths', 8th ESG Conference of Glass Science and Technology, Sunderland, UK, 10-14 September, 2006.

USE AND APPLICATION OF MODELING AND THEIR RELIABILITY

H.P.H. Muysenberg [1], J. Chmelar [2], G. Neff [3]

[1] Glass Service BV, Maastricht, The Netherlands
Adress: Watermolen 22, 6229 PM
Email: e.muysenberg@gsbv.nl
[2] Glass Service Inc., Vsetin, Czech Republic
[3] Glass Service USA Inc., Stewart, USA

ABSTRACT

Mathematical modeling of glass furnaces started around 1965. The question is what can these models do and how reliably are the prediction of such models? The full paper shows validation experiments carried out by several authors over the years. These validations show a fairly good agreement between measurements and models. Certain errors are more likely to come from unknown glass properties and boundary conditions, than from the mathematical model itself. As example we show the error that can be caused when we do not know the glass properties very well.

INTRODUCTION

Modeling has become a powerful tool in recent years with improved algorithms to predict operational conditions in glass melting furnaces, forehearths, regenerators, annealing lehrs, etc. Furthermore, improvements in computer processing speeds have enabled the modeler to work from conventional computers rather than the exotic workstations of the past. However one needs to understand what "it can do" and what "it cannot do". Someone must understand these limits, such that the interpreter is not left with a choice of "Believe it or not".

To date, modeling has been asked to interpret furnace operating conditions, and to predict glass quality parameters. In this regard, models have been used successfully to evaluate alternate furnace designs, as well as to model different operating combinations of combustion processes and electric boosting, or even all electric melters. Additionally, glass quality parameters have also been modeled quite well by evaluations of the glass melting temperatures, and the fining properties of the glass.

Limitations to modeling may include imprecise glass properties at high melting temperatures, or specifically the dynamic furnace processes such as batch charging, or furnace corrosion. Furnaces by their very nature are dynamic processes that are subject to multi-operational influences that cannot always be modeled.

Once you have a close approximation of reality, you can change input conditions whereby the model can predict the revised process trend.

One should not forget that the model is just an approximation of reality, which depends highly on temperature dependent glass properties. If the utilized glass properties are wrong,

55

then the trend prediction will be wrong too. When measuring glass properties such as viscosity, density, electrical conductivity, and thermal conductivity the measurements will contain some errors. This is even more sensitive when measuring thermodynamic properties such as gas solubility or diffusivity where the error can be of an order of magnitude.

The paper will show what modeling can do by showing some examples. Subsequently, we will use a container furnace modeling example and vary the glass properties and show how such variations will change the prediction.

The interest in mathematical modeling is still growing. Especially in recent years it seems to have reached the status of being proven technology. The increased interest in mathematical modeling represents a recognition that modeling can be an effective tool in meeting challenges posed by the problem of glass melting. Glass producers have to reduce costs, satisfy quality requirements, be flexible and meet environmental constraints. The mathematical model can help to predict the effect of the new furnace design on these factors.

Glass Service is the world leader of supplying glass furnace modeling optimizations studies and software licenses.

There is another strong driving force for mathematical modeling, that is the increased computation power. About every year the computation speed, expressed in Million Instructions per Second (MIPS) is doubled. This means that in the last 10 years the power of computers increased by about a factor 1000. This enabled the modelers to model much more complex situations. Today also normal desktop PC's can be used for calculations. As for example today it is normal to calculate the combustion space together with the glass melt. In 1990 the first attempts to do this were made..

The big question for everybody is how reliable is such a model? Can we base our decisions for investment of millions of dollars, on the outcome of such a mathematical model? In this paper we will try to help you to make your own conclusions, if you can believe mathematical models, or not!

FURNACE MODELS IN GLASS INDUSTRY
The paper will focus to modeling of the glass melting process and combustion space from the doghouse till the delivery of glass to the forming process. In figure 1 just the glass flow in the melter is sketched. One important result of modeling is the recirculation or flow of glass in the melter. The continuum process models describe these processes in terms of the well known equations of continuum mechanics (e.g. Navier-Stokes, differential temperature and balances of mass and species) and phenomenological laws describing the relationship between flux and gradients (e.g. Newton's law of viscosity, Fourier's law of heat conduction, Fick's law of diffusion). Of-course the process models must also include mathematical representation of other impacts on flow, heat and mass transfer (e.g. electric heating, bubbling).

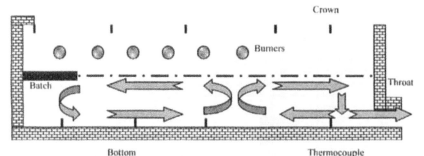

Figure 1. Simplified view of glass flows inside glass melt.

These models are usually divided in primary models, which solve the temperature and flow in the glass and usually a separate model for the combustion atmosphere including combustion and radiation. The secondary model calculates quality, by tracing sand grains, bubbles, stones or in combustion space e.g. NOx or SOx. Figure 2 gives an example of temperature and glass flows in a flat glass melter.

The history of modeling glass furnaces covers around 35 years. The first attempts of modeling melting phenomena by mathematics, in glass furnaces started around 1965 by Mr. W. Trier [1] from the HVG institute in Frankfurt Germany. See also other references for more information [1-12]

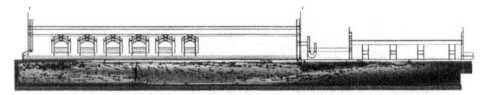

Figure 2. Glass temperature and flow example in a flat glass melter, note recirculation coming back from working end into melter (by Glass Service).

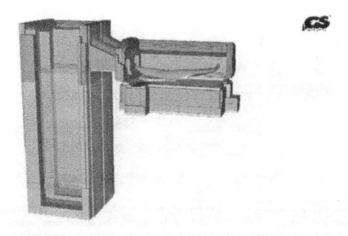

Figure 3. Example of complete modern furnace simulation model including regenerators

VALUE OF MATHEMATICAL MODELS
So as you can see mathematical modeling of glass furnaces has a long history and there are several groups who can help you, or offer you such a service. You probably still have the question, can you trust results of these models and which model should I choose, which one is reliable, what is the difference?

Basically there is or should be almost no difference(s)! As all models are based on the same principle of solving the same equations of mass and energy, we may assume that if one is able to program this without mistakes, the outcome should be the same. But this assumption might be a problem. As you can see Microsoft with professional programmers is not able to program without mistakes. So probably all these technical glass model programmers make mistakes too. Only we do not know if they make a critical mistake or just a minor one. So we need validation. That means comparison of calculated results with known or measured values, can be very helpful to check if there are some serious mistakes. Another problem arises, at temperatures of 1500 °C, we cannot measure so much and what we can measure is not always reliable. We will discuss this in more detail in chapter 4.

The other question is, what is the real value of the outcome of these glass furnace models? This is not only, can I trust that temperature X at spot Y is Z degrees? It means how fast, reproducible and easy can I receive the results?

Let us take the example of a carpenter. The carpenter uses a tool: The hammer. We assume that he has validated and checked this hammer before. But how your house will look like, does not only depend on the quality of the hammer. It depends more on the skills of the carpenter. Does he know the behavior of the wood? Can he listen to you how the house should look like? Can he read your drawings? Can he make you some recommendations based on his experience before starting? How fast can he do it? How nice will the house look after it is finished? The hammer is important, but the final result depends more on his knowledge, experience and skills.

This is also valid for somebody modeling your glass furnace. The mathematical model has to be reliable, but the real outcome depends today more on his knowledge of glass properties and glass furnaces than on the mathematics. To start some modeling, he will have to use some measured glass properties, or if not available take them from his database or even make some estimates. Not all boundary conditions are known exactly. Sometimes he will have to estimate some heat-loss, because the walls are corroded for instance.

So the value of the outcome of the mathematical modeling study depends on:

1. The code (accuracy, coupling error free)
2. Speed of the code (answers this week not after half a year)
3. User friendliness (how fast can one setup a case, also reducing errors)
4. Glass properties (measured, database or estimated)
5. Boundary conditions (measured or estimated)
6. Experience of modeling engineer (does he know that a glass furnace is hot ?)
7. Interpretation of results (not just colorful pictures)
8. Impact on glass melting performance (e.g. quality, bubbles, stones etc..)
9. Post-processing (can you understand pictures, recognize your furnace?)
10. Costs and time.

As you can see the code itself just 1 out of 10 points. But yes, the carpenter is also not able to build your house without the hammer. In this paper we will demonstrate what the effect can be on the relative prediction when we do not now the properties very well.

VALIDATION OF MATHEMATICAL MODELS
So let us see how one can validate mathematical models, if it is possible. The models come up with temperatures in glass melt, refractory, combustion space and for instance exit temperatures of glass and waste gases. Next to it the glass models calculate the speed (flow) of glass in the melter and e.g. working end. Last but not least the real glass furnace produces a certain glass quality or defects, for instance bubble (seed) defects. These bubbles contain certain gases inside which can be checked too.

Temperatures (Error Discussion)
Let see first the temperatures. Temperatures can be measured direct by thermocouples and indirect by optical pyrometers. The direct measurement with the thermocouple can be in the glass or in refractory. Depending on which thermocouple one uses the absolute error can be up to about 10 °C. A typical vertical gradient in the glass melt is about 3 °C/cm. That means the position of the thermocouple is important. It is easy to make an error for the thermocouple position by a few centimeters. The other problem is that the direct measuring thermocouples usually do not last very long and give a wrong value. Derivation could be as high as 50 °C, in some cases 100 °C.

When the thermocouple is inside the refractory its life-time reliability is longer, but the absolute temperature is less reliable. A typical temperature gradient in refractory can be 10 to 20 °C/cm. Thus an error of more than 20 °C is not unlikely.

In the combustion space thermocouples usually receive radiation from all around and do not give a good representative value of the gas around them. Typically a good example of this error we can see for instance just above the regenerator checkers where a thermocouple is exchanging radiation heat with the top of the blocks and usually gives a wrong reading for the preheated air or waste gas. So in the combustion space one should use in fact suction pyrometers.

Also optical pyrometers have some problems. When they measure refractory, one has to set and estimate the emissivity of the refractory material. Maybe the user knows the emissivity as measured in the laboratory, but during operation a glassy layer settled down on the refractory and can influence it. Usually you see that 2 different operators achieve 2 different values.
Measuring the glass (surface) is even more critical and depends highly on the used wavelength of the pyrometer. Depending on the wavelength one measures a certain (small) depth into the glass.

Glass Flow
Than secondly we still have the speed of the glass flow. This is even more difficult than temperature, but still some techniques are available.

Floaters
The first technique is measuring glass surface flows with the aid of floaters. These floaters usually are submerged a little into the glass melt and should follow the speed and direction of the glass surface flow. Of course there are some problems. First problem is to have good access to the furnace to introduce the floaters. Sometimes people also have made holes into the crown to throw them from the top for instance. Secondly we need (good) visual access from several angles at the same time to follow and registrate the position of the floater inside the furnace as function of time. This is very difficult, as usually from each peephole the vision angle is limited and the dimensions are distorted by the lense effect (density difference) of a peephole. One can imagine easily errors of 10-50% for the speed. The third complicated factor is the effect of the flames or combustion gases on the floater and glass surface. If you ever looked to a video of batch at high speed it might be clear to you that in most cases the flames have indeed a large effect (by friction) on the local flow of the glass surface. In an end fired furnace this really can result in a floater floating the opposite direction (pushed by the flames) than the glass flow. This can be easy checked by the effect of the floater after reversing the firing side. So errors of more than 100% can be expected here.

Tracing
The second technique used, is tracing of special components that can be measured/detected in the glass after melting. In the past sometimes some radioactive tracers have been used, but this is today usually not allowed because of safety of employees. Another popular method is for instance addition of some ZnO. This usually has no or a minor effect on glass properties and can be easily detected in the glass sample later. This ZnO is introduced during batch

mixing or directly into the doghouse. Than the ZnO will be melted with the batch and follows the glass currents. Some ZnO will show the fastest or shortest residence time in the furnace and sometimes one even can recognize several peaks identifying several recirculation loops in the furnace. A problem here is mainly the way of introduction or mixing. This can lead to an error of about 0.5 or 1 hour. The other limiting factor is the amount of samples one is willing to analyze, e.g. each 15 minutes during 24 hours. So the resolution is maximum 15 minutes, if the minimum residence time of the furnace is 4 hours than the error can be relative large. The other effect is how long does the glass sample take to come from the end of throat or canal or feeder through the forming process and annealing lehr, because one has to correct for this residence time. This tracing can be easily reproduced, by tracing massless particles in mathematical models. Comparisons are sometimes within half to one hour [21].

Corrosion Profiles
Another method is to check the corrosion profiles on the bottom or side walls. This can confirm flow directions that occur close to the refractory. This was presented in one paper from Chmelar and Schill in 1993 [13]. They not only showed a good agreement with floaters on the glass surface but also between corrosion profiles on the bottom and calculated glass flows.

Measurements and Comparison
Only a few papers have been published showing extensive measurements and comparison modeling results. In the past companies like Corning and Philips had high interest in getting more insight inside their furnaces. Both of them used a special technique to measure glass depth temperature profiles in melters by introducing thermocouples. One introduced thermocouples through the original bottom openings and the other one by introducing them through the original openings in the crown. We will give here some example of results that Philips achieved in cooperation with TNO in one of their TV screens melting furnaces. For more details see the paper of Muijsenberg and Roosmalen [14].

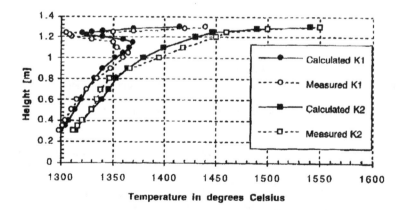

Figure 4. Comparison between calculated and measured glass temperature profile [14].

Figure 4 shows an example of comparison between two measuring points in the beginning of the furnace. K1 is still under the batch and K2 is just after batch melt out. The bottom of the furnace starts at 0.3 meter and the glass surface is at 1.3 meter. So one can see here that locally there can be errors of up to about 30 °C, but the general trend shows a very good agreement between the measurements and calculations.

Chmelar et all [15] shows in contribution from 1997 a comparison between calculated and measured breastwall temperatures of an oxy-fuel furnace. In this case the temperature were measured by an optical pyrometer. These results show in fact the agreement between temperatures in the combustion space as well as the effect of the glass surface temperatures on this breastwall. Besides one peak, which is not shown in the model, the model shows the biggest error near the batch blanket area, up to 40 °C. See figure 5.

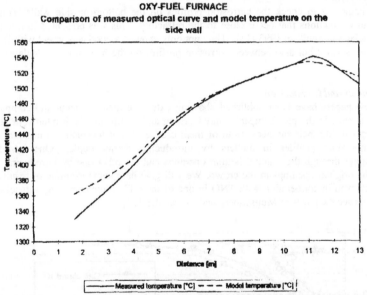

Figure 5. Comparison between calculated and measured side wall temperatures [15].

TECHNICAL COMMITTEE 21 "MODELLING OF GLASS MELTS"

Measurements of velocity and temperature in glass furnaces are not easy and limited. Therefore an alternative method for checking and comparing models was started. Within the International Commission on Glass (ICG) it was decided to start a special Technical committee 21"Modelling of glass melts" to explore the reliability of glass furnace models. This committee was for some time Chaired by Mr. Muschick (Schott glass) and Mr. Muijsenberg (Glass Service) and is now chaired by Mrs. Onsel (Sisecam). Within this committee work it was decided first to model some special defined test case to be modeled by all participants as a round robin comparison. The results of this first test are described in a publication from 1998 [16].

Figure 6. Showing velocity and temperature profiles agreement between different models for defined test case [16]

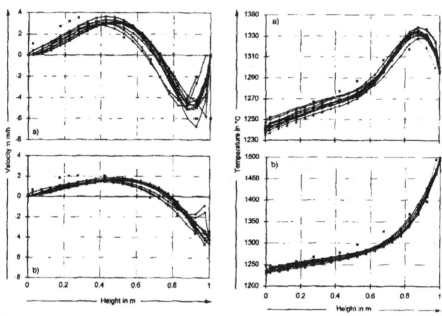

Figures 1a and b. Vertical velocity distribution a) under the bath (at x = 2 m; see figure in table 1), b) in the batch-free region (at x = 5 m; see figure in table 1).

Figures 2a and b. Vertical temperature distribution a) under the batch (at x = 2 m; see figure in table 1), b) in the batch-free region (at x = 5 m; see figure in table 1).

The figure 6 above shows some agreement between different companies using different software to model the same defined case. We can conclude that the general trend in temperature and speed is very similar. But if you look to the absolute results than on a certain selected point you still can find differences between the models of about 10-15 °C. This is probably due to the different approach for different users how to define and setup the boundary conditions. In some cases it also might be due to not good convergence, maybe in a few of the models. If we look for instance to the throat temperature we found a spread of about maximum 15 °C.

Later it was decided to also model an existing furnace with measured data. For this Visteon Glass (Ford) supplied data of the (not longer existing) furnace in Nashville TN. This furnace was a regenerative fired float furnace in which Ford together with Corning measured some glass temperature profiles. As example we show you in figure 7 the results achieved by Glass service with their own Glass Furnace Model.

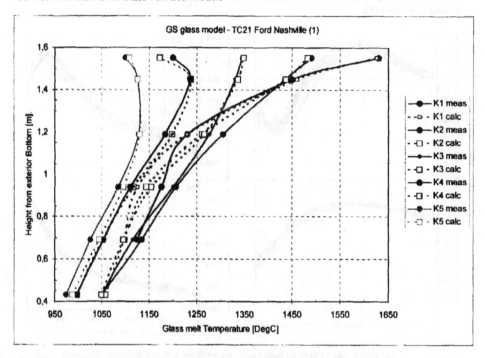

Figure 7. Comparison between measured and calculated temperatures in the former Ford Nashville float glass melter, calculated by GS.

This figure shows a fairly good agreement especially in the trend. Near the bottom and surface, the temperatures are within 10 °C. Note that these results were achieved by a full coupling with the combustion chamber. In the middle of the glass however there can be differences up to about 40 °C, especially in the batch area.

The canal temperature calculated by several participants was within a range from 1110 till 1150 °C. Note that these results were calculated blind, without knowledge of the participants about the results. This shows that in most cases some interaction and fine tuning of (unknown) input data can be necessary. As in this example for instance exact temperatures of preheated air or insulation values of refractory were not known.

Also in Japan such a Round Robin test between several models and one physical model has been carried out. This resulted in good agreements between most participants and the physical model. Chapter 6 will give also such an example.

PHYSICAL MODELING
Another tool to analyze furnaces can be the use of physical modeling. This type of modeling usually consists out of a box made from perspex to be able to watch the flow inside. The box has a geometry that is scaled down according to the real geometry of the furnace and some warm lamps are used to represent the heating. As, in this case, there is really a fluid flowing. some people accept it easier than pure. mathematics. In the past several people also have used this physical modeling to validate the mathematical modeling. I will give some example. One paper was made by Mr Nagao and Wada from NEG [17]. See figure 8 to show agreement between results of a physical fluid experiment with the mathematical model:

Figure 8. Comparison between physical and mathematical model

a) Physical experiment (Taneda,1979)

b) Numerical simulation

Fig. 3 Creeping flow over a fence [Re=0.014]

GLASS QUALITY
What kind of melting efficiency can I expect from this new furnace? How many bubbles per kilogram will it produce? Or how fast can I make a product change? This are the questions a glass producer would like to have answered by the modeling. Mathematical models are a very good tool to help to select the best option, but they are not able (yet) to say exactly how many bubbles per kilogram of glass you will get. This is not only limited by the accuracy of the models, but also due to the fact that we cannot know now, how many bubbles per square meter per time unit will be nucleated. This depends strongly on the applied refractory material and also on how the furnace was constructed and heated up. If for instance the first bottom lining is damaged during heat up and the patch comes into contact with the glass. this can lead

to extensive bubble nucleation. But when the initial modeling was done nobody could expect or foresee this. The good news is that if we assume a certain bubble source and a certain amount of nucleated bubbles, than **Yes** the model can help us to select the best furnace. That means the furnace, which is able to remove most of the bubbles out of the glass before they end up in the product.

This can be done by first calculating the temperature and velocity in the glass melt. Than the redox and gas distribution dissolved in the melt. As last we need to start bubbles from an origin and trace them. During the path travelling through the furnace gases can diffuse into and out of the bubble. For instance oxygen and SO_2 can diffuse into the bubble, make the bubble grow and ascend faster to the glass surface and leave the glass melt. With the following example, executed by Glass Service [18] we want to show you how accurate this kind of prediction can be.

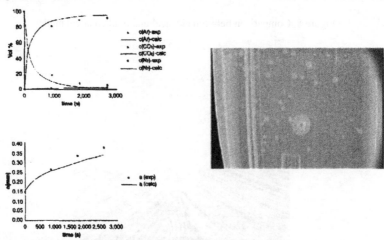

Figure 9. Calculated and measured composition of bubble as function of time [18]

Figure 9 shows us the change of an initial Argon bubble as function of time in soda lime glass. CO_2 (fast) and N_2 (slow) diffuse into the bubble. Consequently the percentages of gases inside and size of the bubble change over time. We can see that the bubble reaches a new equilibrium after 3000 seconds or 50 minutes at 1500 °C. The figure shows the good agreement of the mathematical model and comparison with an experiment and analyses of bubble composition.

In the furnace model we can trace multiple bubbles with varying sources, location and sizes to see if they end up in the product? If they end up in the product we can compare their size and composition with measured compositions from bubbles in the real product. Based on this comparison we can conclude what the possible source of the bubble problem is and try to remove or reduce it.

From the above figure and other experiences we conclude that it is possible to predict bubble compositions and so help to locate the potential source(s).

GLASS PROPERTIES
The biggest difficulty for the glass modeler is given by the glass properties. Most glass producers have the knowledge of the glass composition they are melting, but if you ask for glass properties it is usually limited to density, heat capacity and viscosity. One of the most important data for glass modeling is the thermal conductivity, or in fact radiative properties from e.g. Iron, Chrom and the redox state. This data is not known by the majority of glass producers, because it is not vital information they need for the melting and production of glass.

But today it is possible to measure this property for instance at several institutes like TNO, Glass Service, University in Prague or St. Petersburg [19]. Ofcourse all data should be supplied as function of temperature. The thermal conductivity is usually presented in a polynomial to the third order as function of temperature. The error for this can easily be 10%. Next to it the thermal conductivity in fact varies as function of the local redox state in the glass melt due to the shift of Iron from the 2^+ state to the 3^+ state. Fe^{2+} absorbs heat and Fe^{3+} almost does not.

The modeler needs extensive data for all used properties of the refractories and insulations as function of temperature. That means he needs for instance also density, heat capacity, emissivity and for sure thermal conductivity as function of temperature. For most refractories this is measured and supplied by the manufacturer. But how sure are we that the refractory that is used in a furnace indeed behaves exactly than the one which was measured once some time ago in the laboratory? And what is the effect of corrosion and penetration of some glass into the refractory pores and in-between stones?

If we want to calculate glass quality we also need properties of certain gases in glass like e.g. O_2, CO_2, Ar, SO_2, SO_3, N_2, H_2O. For these gases we need solubility, diffusivity and the redox equilibria as (exponential) function of temperature. Next to it, these properties change as function of glass composition of course.
The availability and reliability of all these glass property data is still limited and is maybe the most important data or part to complete a modeling job successfully [20].

In the following example we will test the effect of a certain error in glass properties on predicted temperatures. Figure 10. Shows the demo furnace we have been using for a test of the effect of glass properties. We have increased and decreased the glass viscosity and thermal conductivity plus and minus 10%. Than we calculated the effect on glass bottom temperatures. With the increased thermal conductivity the temperatures increased 6 °C and with the decreased thermal conductivity the temperature decreased 6 °C.

We wanted than to test what is the change of the trend prediction when we are using the wrong glass properties on a change of 5% fuel increase and bubbling versus no bubbling.

A fuel increase of 5% resulted in the base case into a temperature increase of 51 "C. When we calculated this with the wrong glass properties for viscosity or thermal conductivity the relative prediction was just 50 or 52 "C. So based on these results we may conclude that even if the glass properties are not perfect **that** the simulation models relative prediction is still reliable.

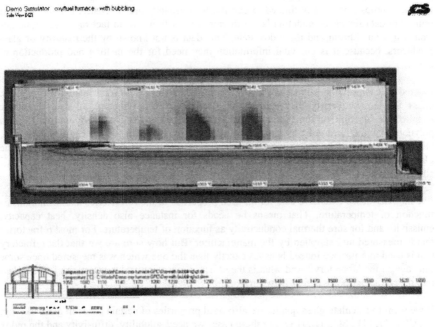

Figure 10. Shows temperature in side view of the center plane of a Demo container furnace simulation. On bottom temperatures are in range of 1300degC.

CONCLUSION
The paper tried to show you the state of the art of glass furnace modeling and its validation. Several mathematical models are available today and probably most of them are validated and accurate. The paper shows that probably experience. glass properties and boundary conditions have a large influence on the outcome and value of modeling. Assuming that thermocouples can show errors. it is fair to expect some agreement between measured and calculated glass temperatures in the range of 10-20 "C. If the difference is larger (at 1 point) than it is more likely that a certain thermocouple is not representative any more, than that the model is wrong. The other explanation of a mismatch can be due to estimated or unknown glass properties. Garbage in results in garbage **out**. Glass furnace models can be a helpful tool to design a better furnace and to optimize the melting performance. It is estimated that today about 1000 furnaces have been optimized/designed by models.

REFERENCES

1. Trier, W. und Stein, A.: Vereinfachtes Rechenmodell des Warmeaustauches durch Strahlung zwischen Flamme und bestimmten Punkten auf der Glasoberflache in einem Glasschmelzwannenofen, Glastechnische Berichte, 38 Jahrgang,, 1965, page 353-361.
2. Hans-Jörg Voss: Mathematisches Modell zur Abschätzung des Energiehaushaltes von Glasschmelzwannenöfen, Glastechnische Berichte, 48 Jahrgang, Heft 9, 1975, page 190-206.
3. Leyens, G.: Beitrag zur Berechnung zweidimensionaler Konvektionsstromungen in kontinuierlich betriebenen Glasschmelzwannen. T. 1 u. 2. Glastechnische Berichte 47, 1974, Nr. 11, S-251-259; Nr. 12 S 261-270.
4. Trier, W.: Temperaturmessungen im Glasbad von Wannenofen. Glastechnische Berichte 26, 1953, S.5-12.
5. Suhas V. Patanker: Numerical Heat Transfer and Fluid Flow, 1980, ISBN 0-07-048740-5
6. L. Nemec, The refining of glassmelts, Glastechnische Berichte, 15, 1974, page 153-159.
7. A. Ungan and R. Viskanta, Three dimensional Numerical modelling of circulation and heat transfer in a glassmelting tank Part 1. Mathematical formulation, Glastechnische Berichte, 60, 1987, page 71-78
8. Simonis F.: Estimation of redox distribution in the melt by numerical modeling. Glastechnische Berichte. 63K, 1990, page 29-38
9. Schill, P.: Calculation of 3-dim glassmelt flow in large furnaces via twogrid method. Glastechnische Berichte. 63K, 1990, page 39-47
10. Schill, P., Chmelar, J.: Bubbles behaviour in the glass melting tank. In: 2nd International Conference "Advances in the Fusion and Processing of Glass", Duesseldorf, 1990.
11. Carvalho, M. da G.; Nogueira, M.: Physically-based modelling of an industrial glass-melting end-port furnace. Glastechnische Berichte 68C2, 1995, page 73-80
12. Muysenberg, H.P.H.: Modeling the combustion chamber of a glass furnace. Presented at 1st Mathematical Seminar on Mathematical Simulation in Glass Melting, Horni Becva (Czech Republic), 1991.
13. Chmelar, J., Schill, P.; Verification of 3D mathematical simulation of glass melting tank, International congres on glass, Madrid,
14. Muysenberg, H.P.H.; Simonis, F.; Roosmalen, R.; Verification of 3D mathematical simulation with measured temperature profiles during furnace operation. Glastechnische Bericht 68C2, 1995, page 55-62
15. Chmelar, J.; Novackova, M.; Safarik, I.; Budik, P.; Mathematical modeling of furnace design. IV International seminar on mathematical simulation in glass melting, 1997, Horni Becva, Czech Republic, page 140-153.
16. Muschick, W.; Muysenberg, E.; Round robin for glass tank models, Report of the International Commission on Glass (ICG), Technical committee 21"Modelling of glass melts", Glastechnische Berichte. Glass Sci. Technol. 71no6, 1998, page 153-156
17. Nagao, H.; Wada, M.; Three dimensional numerical simulation for visualization of fluid flow, 1990 ?.
18. Ulrich, J.; Nemec, L.; Matyas, J.; Mathematical modelling for the identification of defect bubble sources, International glass review, 1998.
19. Endrys J.; Measurements of radiative and effective thermal conductivity of Glass Service, IV. International seminar on mathematical simulation in glass melting, 1997, Horni Becva, Czech Republic.
20. Klouzek, J.; Determination of the equilibrium partial pressures of sulphur dioxide and oxygen in float glass melt, III. International seminar on mathematical simulation in glass melting, 1995, Horni Becva, Czech Republic.

21. Hermans J.M.; Roosmalen M.P.W.; A tracer trial on a TV-Funnel tank, Proceedings, ICG Annual meeting 2000, Amsterdam.
22. Choudary, M.; A Three-Dimensional Mathematical Model for Flow and Heat Transfer in Electrical Glass Furnaces, IEEE Transactions on Industry applications, Vol 1A-22, No5, September/October 1986.
23. Choudary, M.; Norman, T.; Mathematical modeling in the glass industry: An overview of status and needs, Glastechnische Berichte, Glass Sci. Technol. 70 no. 12, 1997, page 363-370
24. Murmane, R.A.; Johnson, W.W.; Moreland, N.J.; The analysis of glass melting processes using three-dimensional finite elements, International journal for numerical methods in fluids, Vol 8, 1491-1511, 1988.
25. Mase, H.; Oda, K.; Mathematical model of glass tank furnace with batch melting process, Journal of Non-Crystalline Solids 38&39, 1980, 807-812.
26. Hayes, R.; Wang, J.; Mcquay, M.; Webb, B.; Huber, A.; Predicted and measured glass surface temperatures in an industrial regeneratively gas-fired flat glass furnace.

ENVIRONMENT

AIR EMISSION REQUIREMENTS - PAST, PRESENT AND FUTURE

C. Philip Ross
GICI

ABSTRACT
Historically, the glass industry has been faced with gradually more stringent air emission control regulations. This paper will provide an overview covering the evolution of regulations for glass furnaces, key issues associated with controlling the current criteria pollutants, and an understanding of near future control requirements.

INTRODUCTION
The process of converting industrial minerals to molten glass in high temperature furnaces has certain inherent gaseous and solid particulate emissions into the earth atmosphere. The growing awareness of concerns with air emissions resulting in health concerns has led to an ongoing evolution of more stringent regulations to limit emissions from most industrial processes, including glass melting. From a regulatory perspective, most legally imposed regulations have significantly impacted the glass industry over the past five decades.

The "Pollution Prevention and Abatement Handbook" defines Air Pollutants as: "Any substance in air which could, if in high enough concentration, harm man, other animals, vegetation, or material. Pollutants may include almost any natural or artificial composition of matter capable of being airborne. They may be in the form of solid particles, liquid droplets, gases, or in combinations of these forms. Generally, they fall into two main groups: (1) those emitted directly from identifiable sources and (2) those produced in the air by interaction between two or more primary pollutants, or by reaction with normal atmospheric constituents, with or without photo-activation. Exclusive of pollen, fog and dust, which are of natural origin, about 100 contaminants have been identified and fall into the following categories: solids, sulfur compounds, volatile organic chemicals, nitrogen compounds, oxygen compounds, halogen compounds, radioactive compounds, and odors."

To varying degrees, the following categories of air pollutants are potentially emitted from a high temperature glass melting process and may exist in the "stack" exhaust gases.

Products of Combustion Gases

NO_X - An inherent by product where Nitrogen is present within the combustion process
CO - A remnant of incomplete combustion of a carbon containing fuel
SO_X - An oxidized gas by product of Sulfur in the fuel

Batch Raw Material Ingredient Evolution

SO_X - The evolution product of sulfur containing raw materials, typically from the refining process
Condensable Particulate - Compounds created within the cooling exhaust process of gases evolved from the melt (including Alkali, Borates, Sulfates, and Carbonates)
Heavy Metals - Compounds of low vapor pressure metals (such as Arsenic, Lead, Selenium, etc.)

Solid Dust - Very fine particles of raw materials entrained on the exhaust gases

Refractory Sourced Emissions

Chrome Compounds - Chromic Oxide refractories having reacted with glass making raw materials

The methods actually employed for reaching compliance with environmental regulations will depend upon many factors - including specific furnace design, site constraints, relative capital investment vs. operating costs, risks relating to labor and maintenance requirements, and the potential impact upon glass quality or production efficiencies. Some of the control technologies being employed on glass furnaces include -

NO_X - Process Modifications, Oxy Fuel, Ammonia Injection , Air Staging, Gas Reburn
SO_X - Avoid Sulfur containing Fuels, Batch Sulfate alternatives , Add-on Scrubbers /
Reactors
Particulate - Add-ons (Bag Houses, Electrostatic Precipitators, Scrubbers)

PARTICULATE ISSUES

Historically, glass furnaces were first regulated for particulate. Visual emissions by stack opacity were typically restricted to less than 20 % ("#1 Ringlemann" equivalent). For Soda-Lime glasses, the predominant particulate chemistry is sodium sulfate, while alkali borate particulate occurs from Fiber Glass and other Borosilicate glasses. Common process modifications and best practices, adopted by furnace operators to minimize condensable particulate emissions, have historically included:

- Reductions in Batch Sulfate levels and optimization of the refining process
- Reformulating of Borosilicate glasses to reduce or eliminate B_2O_3
- Effective Batch Wetting and raw material particle size optimization in the batch charge
- Higher Cullet content, resulting in less Sulfate per ton of glass melted
- Electric Boosting to limit melt surface temperatures and sustain required pull rates
- Furnace Design configurations to avoid aeration of batch particles, promote glazing of the initial batch charge surface, and to reduce gaseous velocities over the melt surface
- Conversion from Fossil Fuel Oil combustion to no ash, low sulfur fuels (natural gas)
- All- Electric Melting

Most operating permits incorporated maximum pull rates, maximum firing rates, maximum temperatures and other quantified limits based upon the above mentioned process variables; and often limited to values occurring during compliance demonstration testing. State and regional authorities (New Jersey and California) began setting mass emission limits for particulate in the early 1970's.

Any facility that commences *construction or modification* after June 15, 1979, is subject to the requirements of the Federal New Source Performance Standards. - *40 CFR 60.292 Standards for Particulate Matter*. Limits are expressed in grams of particulate per kilogram of glass melted, and vary for "New" vs. "Modified" furnaces, as well as the type of fuel. Most furnaces also have visual opacity limits significantly below 20 %, with a limit often based upon actual measurements taken during compliance demonstration test periods.

The requirement of compliance with the Federal NSPS and appropriate permitting for a furnace is typically triggered by one of the following conditions.

- Construction of a new facility after 6/15/79
- Any physical or operational change to an existing facility resulting in an increase in emission rate
- A designated unit considered a "replacement" for an existing unit
- A "reconstructed" facility costing > 50 % of the "fixed capital cost" of a comparable new facility

"New Source" standards are more restrictive than those for a "Modified Source". Existing sources converted to Oxy-Fuel have been considered by some agencies as a "process modification", if their is no increase in Particulate emissions.

Specific definitions are used to apply the rule's requirements, including:

"Glass melting furnace" means a unit comprising a refractory vessel including foundations, superstructure and retaining walls, raw material charger systems, heat exhaust, melter cooling system, exhaust system, refractory brick work, fuel supply and electrical boosting equipment, integral control systems and instrumentation, and appendages for conditioning and distributing molten glass to forming apparatuses.

"Rebricking" means cold replacement of damaged or worn refractory parts of the glass melting furnace. Rebricking includes replacement of the refractories comprising the bottom, sidewalls, or roof of the melting vessel; replacement of refractory work in heat exchanger; replacement of refractory portion of the glass conditioning and distribution system.

SO_X ISSUES

Other than Los Angeles, there are no SO_X emission limits stringent enough to impact natural gas fired glass furnaces in the U.S. The primary source of SO_X emissions come from Sulfates in the batch, which are the major contributor of Particulate in Soda-Lime glasses. Present practice for Soda-Lime is to use low sulfate batches, and consequently SO_X emissions are less than 1.0 gm / kg. (2 lbs. per ton of glass melted).

NO_X ISSUES

Southern California's SCAQMD Rule 1117 set the first significant NO_X limits in the country. Prior to 1987 the container glass industry had historically uncontrolled NO_X emissions from conventional regenerative container glass melting furnaces in the range of 8-10 lbs/ton of glass pulled. Float furnaces often have NO_X emissions in excess of 20 lbs. / ton. Rule 1117 (Emissions of Oxides of Nitrogen from Glass Melting) called for significant reductions in NO_X emissions in 1987 to 5.5 lbs/ton and 4.0 lbs / ton in 1993. Other agencies have set similar limits for Container Glass. California's San Joaquin Valley's NO_X Rule 4354 now restricts Float Glass furnaces to 7.0 lb. / ton.

Process Modifications have been investigated and pursued since the California Air Resources Board (CARB) Model Rule was developed in the early 1980's. Since 1987, most of the practical options listed below were implemented by existing facilities to meet Los Angeles' original Rule # 1117.

- Excess O_2 control through Mass Flow Ratio control and continuous O_2 measuring sensors
- Higher levels of Electric Boost to lower combustion zone temperatures
- Increased Cullet additions for reduced Gas firing rates
- Sealed, Low Velocity Burner systems integrated into Port design changes for flame shaping

Each furnace and facility have unique, site specific differences. To meet compliance with specific NO_X limits regulation, the following strategies have been utilized on air fired furnaces:

- Oversized melter with heavy electric boost, high cullet and lower pull rates
- End-Port firing on Oil with added Scrubber & Bag House for SO_X and Particulate Control
- Converted Side Port to Large End Port, using Under Port Sealed Burner System
- Low Bridgewall Temperature and heavy boost
- Addition of Oxy-gas boost burners
- Furnaces using Nitrates in the batch are limited to ~ 7 lb. / ton
- Ammonia Injection
- Sorg LoNOx Furnace
- Batch / Cullet preheating to lower energy input

In the past, glass furnaces had essentially no CO limits (typical furnaces were less than 20 ppmv, unless combustion modifications were being used for NO_X control). The latest standard in California imposes a 30 ppmv limit in the San Juaquin Valley with Rule 4353.

CAAA 1990

The Federal Clean Air Act Amendment of 1990 established ambient air quality standards for Ozone. Regions not in compliance with the standard are required to regulate NO_X emitters (including glass furnaces). Areas with worse air are setting more stringent standards.

Title I addresses urban air quality problems in non-attainment areas. Three air pollution problems are covered: smog /ozone, caused by nitrogen oxides (NO_X) and volatile organic compounds (VOC's); carbon monoxide (CO); and particulate matter. Glass manufacturing is listed as a category of sources that contribute to non-attainment of the national ambient air quality standard for PM_{10} (particulate matter less than 10 microns) and $PM_{2.5}$ (less than 2.5 microns). In reality, most condensable particulate from glass furnaces is less than 2.5 microns.

The EPA has classifications for smog /ozone include *marginal, moderate, serious, severe,* and *extreme.* Depending on an area's air pollution severity, the EPA enforces regulations for specify different air pollution control limits and plant's must implement appropriate control measures.

Title III seeks to control 189 *Hazardous Air Pollutants* (HAPs, also called air toxins) that are hazardous to human health or the environment. HAPs are typically carcinogens, mutagens, and reproductive toxins.

Major sources are defined as emitting 10 tons / yr. or more of any HAP or 25 tons / yr. of any combination of HAPs. EPA indicates it may decrease the emissions level required for classification as a major source from 10 tons / yr. to tons / yr. or even 0.1 tons / yr. These will require the EPA to issue control standards, called *Maximum Achievable Control Technology* (MACT) standards, for each source category.

The standards will be based on the best demonstrated control technology or practices within a specific industry. Different standards will apply to existing and new sources. If a plant is an existing source, the MACT standard can be a control technique that is at least as stringent as the average of the cleanest 12 percent of sources in the same industry. But if it is a new source, the control technique can be no less stringent than the best-controlled existing major source.

Title V enacts a national operating permits program for any major source subject to Title I or Title III and must obtain an operating permit to ensure the plant complies with the applicable requirements. "Major Sources" requiring permits are defined as having the potential to emit 10 tons per year of a single or 25 tons of combined hazardous air pollutants. The permitting procedure better defines emission inventories to be used for setting emission caps or forced reductions in non attainment areas. Monitoring requirements will be necessary to identify "periods of noncompliance." Enforcement can include civil, as well as criminal penalties.

The 1990 CAAA established Federal requirements for BACT, RACT, BARCT, LAER. Local and Regional Districts are allowed to perform a structured process to construe what limits and technologies meet these categories, for various emission sources. District or State air quality attainment plans must be designed to achieve and maintain ambient air quality standards by the earliest practicable date, and include regulations which require control technologies for existing and new sources.

RACT is defined in 40 CFR section 51.100(o) as follows:

> "Reasonably Available Control Technology means devices, systems, process modifications, or other apparatus or techniques that are reasonably available taking into account (1) the necessity of imposing such controls in order to attain and maintain a national ambient air quality standard, (2) the social, environmental, and economic impact of such controls, and (3) alternative means of providing for attainment and maintenance of such standard..."

RACT is required in plans for all districts designated as "Moderate". RACT should be the most stringent of the following control options:

- The most effective emission limits in existing regulations that are currently in effect in any district whose non-attainment status is designated as moderate.
- Emission limits identified in existing Suggested Control Measures (SCMs), model rules, EPA's Control Techniques Guidelines (CTGs) or other such documents.
- The lowest emission limit that can be achieved by the specific source by the application of control technology taking into account environmental impacts, technological

feasibility, cost-effectiveness, and the specific design features or extent of necessary modifications to the source.

- The lowest emission limit achieved for the source category that is technically feasible, economically reasonable or achieved in practice anywhere (including outside of the U.S.).
- Any combination of control technologies that will achieve emission reductions equivalent to that resulting from the most stringent option listed above.

The application of BARCT (Best Available Retrofit Control Technology) will be required for districts that are designated as either "serious" or "severe". BARCT is generally defined as "an emission limitation that is based on the maximum degree of reduction achievable, taking into account environmental, energy, and economic impacts by each class or category of source."

BARCT should be the most stringent and cost-effective of the following control options:

- The most effective limits in effect in any in the U.S., or in any other country for that source category.
- The most effective limit for a source category determined; to a reasonable degree of certainty, to be achievable in the near future.
- Any combination of control technologies that will achieve emission reductions equivalent to that resulting from the most stringent option listed.

The process of developing a definition for BARCT involves a structured process, including a "top down" cost effectiveness analysis for BARCT determinations. It is always preferred that the glass manufacturers actively participate in this defining process. All applicable control measures (i.e., add-on controls, process modifications, alternate fuels, etc.) for applicable source category (ies) are ranked from highest to lowest emission reduction of non-attainment pollutant (s). For the remaining control measures, a second ranking from best to worst cost-effectiveness is created.

CAAA 1996

On December 31, 2002 the US EPA published its final revisions to the New Source Review (NSR) programs mandated for both attainment and non-attainment areas. These revisions include Baseline Emissions Determinations, Actual to Future-Actual Methodology, Plant wide Applicability Limitations, and Pollution Control Projects.

Looking ahead, the Glass Industry's most viable option for long term compliance for strict NO_X limits will probably be conversion to 100% oxygen combustion. All glass industry segments have successfully converted to Oxy-Fuel. The implementation of this technology for meeting future environmental compliance will initiate a significant driving force to integrate waste heat recovery schemes - such as batch / cullet preheating, cogeneration, or gas reformer technology.

RECLAIM

The trading of emission reduction credits has been a reality for glass furnaces in the U.S. since 1994. The South Coast Air Quality Management District (SCAQMD) in Los Angeles developed a market based regulatory program called the Regional Clean Air Incentives Market (RECLAIM) program. Traditional regulations, known as command-and-control, had previously set specific limits on each piece of equipment and each process that contributes to air pollution. RECLAIM encloses the facility in an imaginary "bubble." Rather than regulating each source, SCAQMD regulates the total pollution in the bubble, and lets businesses decide what equipment, processes and materials they will use to meet their emission limits. Under RECLAIM, these allowable emission limits decline a specific amount each year. Companies are free to choose the most cost-effective, economical ways to reduce pollution and operate within their allocation.

Participants in RECLAIM receive trading credits equal to its annual emissions limit. Credits are assigned based on past peak production and the requirements of existing rules and control measures. Credits are assigned each year and can be bought or sold for use within that year. Facilities must hold credits equal to their actual emissions.

The RECLAIM program applies to <u>stationary sources</u> that emit four or more tons per year from permitted equipment. It required industries and businesses to cut their emissions by a specific amount each year. The program targeted a 70% reduction for nitrogen oxides (NO_X) and a 60% reduction for sulfur oxides (SO_X)" over a nine year period. NO_X was included in the program because it is a precursor to ozone, for which the District is a Federal "extreme non-attainment" area. SO_X is included in the program because it is a precursor for fine particulate matter (PM_{10}), for which the Basin is in "serious" non-attainment.

Businesses that beat their reduction targets can trade their credits on the open market. Using market forces allows pollution to be cut in the most economical way. To monitor emissions at larger sources, RECLAIM requires use of continuous emission monitoring systems to determine actual mass emissions from these sources. These emissions are electronically reported to the District on a daily basis. The sale of credits by over control technologies (such as Oxy-gas for NO_X and scrubbers for SO_X) has yielded glass manufacturers significant revenues.

PSD / NSR

Under the original 1977 clean air act (40 CFR 52.21), Congress established Prevention of Significant Deterioration (PSD) program (applicable in areas attaining national ambient air quality standards) and non attainment New Source Review (NSR) program (applicable in areas not attaining such standards). "Major sources" and "major modifications" of criteria pollutants (O_3, SO_2, etc.) and "modifications" must be permitted under PSD and/or NSR programs

"Major modification" is defined as a "physical change or change in method of operation" of a major source that "would result in a significant net emissions increase" of any regulated pollutant. EPA's regulations include two significant exclusions from applicability: 1) activities that constitute "routine maintenance, repair and replacement" and 2) emissions increases attributable to an "increase in hours of operation or production rate".

If a physical or operational change constitutes a major modification, a source must install "Best Available Control Technology" (BACT) under PSD and "lowest achievable emission rate" (LAER) technology under non-attainment NSR; it also must comply with other requirements (including obtaining offsets for emissions increases in non-attainment areas)

EPA enforcement officials have concerns that some major facilities may have failed to comply with PSD/NSR permitting requirements since the enactment of these regulations. EPA is seeking installation of emission controls on sources which failed to obtain PSD/NSR pre construction permits for "major modifications". Investigation begins with the issuance of "Section 114" request letters, followed by subsequent requests for information, inspections, and issuance of notices of violation (NOV's). Section 114 requests seek information on all plant changes since 1978 or 1980; but for the glass industry, the specific focus is on furnace rebuilds.

Under EPA's rules, activities that constitute "routine maintenance, repair and replacement" are excluded from the meaning of "physical change" and thus are exempted from NSR applicability. EPA has confirmed in the past that rebricking is routine for glass industry; e.g., in 1980 final NSPS preamble, EPA explained that "the rebricking exemption was not questioned due to the regularity and necessity of the operation to this industry;" in 1994 draft NSR rule preamble, EPA listed, as an illustration of routine maintenance, *"furnace refractory maintenance, repair or replacement with new refractory material at . . . glass facilities"*

EPA enforcement statements now indicate EPA's desire to limit scope of exclusion to "frequent, traditional and comparatively inexpensive" activities. EPA's current position seems to be that any repair or replacement activity that results in efficiency improvements or, even where it does not, that is not "comparatively inexpensive" does not come within the "routine" exclusion.

When furnace rebuild projects involves physical or operational change, the principal determinant in assessing NSR applicability is whether change results in a "net emissions increase". This is being defined as "<u>any</u> increase in actual emissions" from the change. Determining whether a net emissions increase occurs hinges on how post-change emissions are to be calculated.

EPA's NSR rules provide that "actual emissions" shall be equal to the "potential to emit" of a unit where, and only where, it "has not begun normal operations". EPA's methodology for determining "an increase in actual emissions" has changed dramatically over time. In early 1980's, after rule was promulgated, EPA recognized that it should compare pre-change to post-change actual emissions and only applied the "actual-to-potential" test where unit had truly not begun normal operations, i.e., there is no operating history. In late 1980's, EPA had begun to assert that "actual-to-potential" test generally should be applied.

In the future, EPA's enforcement initiative is likely to be a high priority. There is still uncertainty regarding the scope of the "routine" exclusion and how increases in "actual emissions" are to be determined. This uncertainty about PSD / NSR applicability may continue to lead some companies to enter into settlements with EPA.

CAIR BACKGROUND

On September 24, 1998, EPA finalized a rule (known as the NOx SIP Call) requiring 22 States and the District of Columbia to submit State implementation plans that address the regional transport of ground-level ozone. The intent of these plans is to decrease the transport of ozone across State boundaries in the Eastern half of the United States, particularly emissions of nitrogen oxides (a precursor to ozone formation). The NOx SIP call builds upon analyses conducted by the Ozone Transport Assessment Group (OTAG).

On March 10, 2005, EPA issued the Clean Air Interstate Rule (CAIR). The CAIR requires certain upwind States to reduce emissions of nitrogen oxides (NOx) and/or sulfur dioxide (SO_2) that significantly contribute to non attainment of, or interfere with maintenance by, downwind

States with respect to the fine particle and/or 8-hour ozone national ambient air quality standards (NAAQS). The CAIR requires these upwind States to revise their State implementation plans (SIPs) to include control measures to reduce emissions of SO_2 and/or NOx. Sulfur dioxide is a precursor to $PM_{2.5}$ formation and NOx is a precursor to $PM_{2.5}$ and ozone formation.

On May 12, 2005, the EPA published the final "Rule to Reduce Interstate Transport of Fine Particulate Matter and Ozone" (Clean Air Interstate Rule or CAIR) (70 FR 25162). In this action, EPA found that 28 States and the District of Columbia contribute significantly to non attainment of, and interfere with maintenance by, downwind States with respect to the NAAQS for fine particles ($PM_{2.5}$) and/or 8-hour ozone.

The EPA conducted extensive air modeling to determine the extent to which emissions from certain upwind States were impacting downwind non attainment areas. All States found to contribute significantly to downwind $PM_{2.5}$ non attainment are included in the CAIR region for $PM_{2.5}$ and are required to reduce annual emissions of SO_2 and NOx. All States found to contribute significantly to downwind 8-hour ozone non attainment are included in the CAIR region for ozone and are required to reduce NOx emissions during the 5-month ozone season (May-September).

The first phase of NOx reductions starts in 2009 (covering 2009-2014) and the first phase of SO_2 reductions starts in 2010 (covering 2010-2014). The second phase of both SO_2 and NOx reductions starts in 2015 (covering 2015 and thereafter). Each State covered by CAIR may independently determine which emission sources to control, and which control measures to adopt.

OTC BACKGROUND

The (OTC) is a multi-state organization created under the Clean Air Act (CAA). Their responsibility includes advising EPA on transport issues and for developing and implementing regional solutions to the ground-level ozone problem in the Northeast and Mid-Atlantic regions. OTC members include: Connecticut, Delaware, the District of Columbia, Maine, Maryland, Massachusetts, New Hampshire, New Jersey, New York, Pennsylvania, Rhode Island, Vermont, and Virginia.

The Coalition of Northeastern Governors (CONEG) believes that particular attention must be paid to improving air quality to protect public health and the environment, equitable environmental requirements should apply across the country, and one region should not adversely impact another. CONEG supports the ongoing efforts of the Ozone Transport Assessment Group (OTAG).

Because ambient air quality in this region is often influenced by emissions from "upwind" states, the Commission believes Model Rules must be developed and implemented for the entire Region. The Committee has compiled a list of Reasonably Available Control Technology (RACT) Control Technique Guidelines (CTGs) categories that should be updated. The OTC Model Rules have been recommended as a logical starting point for RACT updates for State adoption in future rule making.

The OTC's Control Strategy Committee has recommended NO_X control measures for 33 industrial processes, including Glass. These recommendations are in final draft form, and the last opportunity for modifications will be by written submission to OTC or during comment periods at their next meeting (Nov. 15, 2006). At this meeting the Commission expects to approve the model control measures and recommend their member states to adopt them into new emission regulations.

The driver for these model rules are ambient air computer models showing that there must be significant reductions of existing emission inventories to meet Clean Air Act standards. To accomplish this, OTC is using BACT criteria for selecting technologies which have the highest percentage of reduction and have cost effectiveness numbers no greater than ~ \$2,500 per ton of NO_X reduced. CEMS will be required for compliance assurance.

The Glass Furnace NO_X emission inventory in the OTC Region is currently estimated to be approximately 15,000 tons per year (~ 5 % of the total emissions inventory). At the time of this writing, the OTC is considering a change from their Model Rule requiring Oxy-Fuel melting, to an emission limit rule - similar to the SJVUAPCD Rule 4354. Such a rule would reduce glass furnace emissions by ~ 44 %.

A very strong message coming out of the OTC region is that the Federal EPA needs to revise emission limits on a broader (national) basis, such as what is being done under CAIR for Electrical Generating Units (EGU's). Many regional ambient air quality problems are affected from "up wind" sources out of their regulatory authority.

OSHA ISSUES ON HEXAVALENT CHROMIUM

In April 2003, a U.S. Court of Appeals ordered the Occupational Safety and Health Administration (OSHA) to promulgate a standard governing workplace exposure to hexavalent chromium. OSHA has published a final standard for occupational exposure to hexavalent chromium in the Feb. 28, 2006, Federal Register.

The new standard lowered OSHA's permissible exposure limit (PEL) for hexavalent chromium, and for all Cr (VI) compounds, from 52 to 5 micrograms of Cr (VI) per cubic meter of air as an 8-hour time- weighted average. The standard also includes provisions relating to preferred methods for controlling exposure, respiratory protection, protective work clothing and equipment, hygiene areas and practices, medical surveillance, hazard communication and record keeping.

Glass manufacturers have an obligation to measure the level of hexavalent chromium in all areas of their facilities which could expose their employees to excessive levels. Sources of hexavalent chromium can include exhaust gases from furnaces or forehearths utilizing Chrome containing refractories, Chrome containing glass colorants, or operations where hexavalent chromium may be volatilized (such as color and traditional forehearths). Engineering controls will be preferred over personal respirator equipment.

FUTURE ISSUES

Since 1990, additional regulations have been applied to "toxic" particulate - particularly Lead and other heavy metals (Arsenic, Selenium, Cadmium and Chrome compounds). At the time of this presentation in 2006, there are continuing regulatory initiatives to further limit emissions from glass melting furnaces.

The Clean Air Act Amendment requires the EPA to address the reduction of Urban Air Toxics. Section 112(b) lists 189 Air HAP's, but 33 specific Urban HAP's (the "dirty thirty") are to be specifically addressed in new Area Source Standard regulations for the emitting industries. Their authority for these actions come from the CAA's Sections 112(c)(3) and 112 (k)(3). Facilities which include toxic compounds among raw materials for glass melting will be subject to the new standards. Seventy area source categories have been listed for standards development. Standards for 15 are complete, 5 have consent decree dates for promulgation, and 50 are subject to ongoing negotiation to establish consent decree dates.

Specific materials potentially subject to the new regulations for glass include compounds of Arsenic, Cadmium, Chromium, Lead, Manganese, Mercury, and Nickel. Area source standards are to be technology-based and capable of reducing emissions by ~ 90 %. EPA currently is planning to propose an equipment standard. EPA is considering requirements to properly install, maintain, monitor and keep records on the performance of air pollution control devices on the processes subject to the regulation. EPA is currently considering emission controls on raw material handling and processing lines and on furnaces. There could also be consideration for controls if Chrome containing refractories are used in the furnace.

Section 112 of the Clean Air Act (CAA) requires the development of standards for area sources which account for 90% of the emissions in urban areas of the 33 urban hazardous air pollutants (HAP) listed in the Integrated Urban Air Toxics Strategy. At this time the EPA is attempting to quantify the extent of potential emissions of HAP's from Glass manufacturing. They have been using existing Permits, State Inventories and the federal Toxic Release Inventory (TRI). To date they have values for ~ 170 facilities, but believe there are as many as 500 glass related sources.

A recent Draft by EPA for National Emission Standards for Hazardous Air Pollutants (NESHAP) will require reporting by glass manufacturers as to the weight of "hazardous" materials in their batch. If the level exceeds 1% of the batch weight or use exceeds 20 tons / year, stack source tests will be required to quantify the level of emissions. If defined levels are exceed, the use add on controls such as a bag house or an electrostatic precipitator (ESP) will be required.

EPA's greatest concern seems to be with the use of Arsenic, and other colorants used in the Press & Blown and tableware industry (Cadmium, Lead, and Antimony). For container glass, the major issue will involve colorants (Selenium, Chromium, Nickel, Cobalt). To date, EPA has found limited contacts in industry to discuss these issued.

GREENHOUSE GAS ISSUES

Similar to European initiatives, CO_2 emissions may well be Federally regulated under pressure from "Global Warming" concerns. On August 31, 2006, California's legislature approved the broadest restrictions on carbon dioxide emissions in the nation. The California bill requires a 25 percent cut in carbon dioxide pollution produced within the state's borders by 2020 in order to bring the total down to 1990 levels. In at least eight other states, political momentum is building to take similar steps to limit emissions of greenhouse gases linked to climate change, a trend that could increase the pressure for a national system.

The California legislation also provides a statewide market system designed to make it easier for heavily polluting industries to meet the new limits. They would be able to buy "credits" from companies that emit lower emissions than the caps allow, rather than having to invest in lower greenhouse gas emitting technologies.

It is still unclear how this legislation will effect the glass industry, CO_2 emissions from glass melting include the results of natural gas combustion, as well as evolution from raw material Carbonates. For Soda-Lime glasses, there are typically more than 35 tons of CO_2 emitted per 100 tons of glass melted. Recycling of 6 tons of cullet reduces 1 ton of CO_2.

CONCLUSIONS

• Regulations will continue to become more stringent in Non-attainment Regions

• Agencies will expect newer technologies to be more efficient than what are currently available

• Future furnace Types and Designs will give greater consideration for emission compliance

• Compliance will have higher priorities in operations

• More CEMS will be required for Compliance Assurance

• New Source Requirements will become more stringent

DRY SORBENT INJECTION OF TRONA FOR SO$_x$ MITIGATION

John Maziuk
Technical Development Manager
Solvay Chemicals

ABSTRACT

Boilers, furnaces, metals and ceramics manufacture, and similar processes produce waste off-gases containing acidic species that, if released uncontrolled, are harmful to human health and the environment. The acid gases of primary concern are oxides of sulfur (SO$_X$) and hydrogen halides (HX). Gas scrubbing with alkaline solutions and solids is employed to reduce emissions of these acid gases. Different scrubbing technologies have been practiced for decades but least expensive to implement and simplest to operate are usually "dry scrubbing" or dry sorbent injection (DSI) systems. In DSI systems a dry powdered alkaline material is injected into the hot gas stream to neutralize the acidic species, and the resulting solid salts and remaining excess alkaline material is collected by a downstream particulate capture device. These systems are capable of high levels of acid gas reduction. Various alkaline materials, both chemically processed and naturally occurring, have seen application in dry scrubbing. Dry hydrated lime, a calcium based alkaline sorbent, is in wide use in dry scrubbing, largely because of its moderate cost and availability. However, sodium based alkaline sorbents such as sodium bicarbonate, nacholite, and trona can provide particular advantages that make them a superior choice in certain applications. A mechanically refined trona in both a powder and granular form is being used throughout these industries for SO$_x$ mitigation and recently at a few glass production plants with good success and having economic advantages over other sorbents.. This paper will discuss the application process, important variables to insure optimal sorbent usage, the chemistry of the operation and data collected under various conditions including glass production. This paper summarizes the advantages of DSI systems and of dry injection of sodium based alkaline sorbents in particular.

INTRODUCTION

Fossil and waste fuel fired boilers and incinerators, primary and secondary metallurgical production processes, petroleum refinery catalyst regeneration, foundry operations, brick kilns, and other furnaces associated with production of various mineral products are common industrial sources of SO$_X$ and HX are combustion or other high-temperature processes involving materials containing sulfur and/or halides.

When materials being processed in oxidizing atmospheres contain sulfur, the off-gases contain sulfur dioxide (SO$_2$) and small quantities of sulfur trioxide (SO$_3$) arising from excess O$_2$ and catalysis. The SO$_2$ is present as a gas and the SO$_3$, which hydrolyzes with moisture in the gas stream, is usually present as an aerosol (sulfuric acid mist). When the materials being processed contain halides, such as fluorine, chlorine, or bromine, the off-gases will contain HF, HCl, or HBr. As with SO$_3$ these gases will also hydrolyze with available moisture to form acid aerosols.

SO$_2$ is a criteria pollutant that effects both public health and the natural environment, and as such it has both primary and secondary National Ambient Air Quality Standards (NAAQS). In addition, Title IV of the 1990 Clean Air Act Amendments mandates reductions in SO$_2$ emissions from electric utilities.

DISCUSSION

Although technologies exist for "cleaning" fuels or other materials before processing to remove acid gas precursors (sulfur and halides) and thereby limit acid gas emissions, pre-treatment is generally not economically viable. The release of acid gases to the environment is most often controlled by "scrubbing" the waste gases from the processes prior to release through a stack or other process vent.

Acid Gas Scrubbing Technologies

"Acid gas scrubbing technologies are usually based on multi-phase acid-base neutralization reactions. Neutralization reactions are typically fast and exothermic, and equilibrium conversions are high. However, in practice acid gas scrubbing neutralization reaction rates are mass-transfer limited and equilibrium conversions are not reached because the processes are continuous and available reaction times are limited. The effective rate of reaction and thereby the extent of acid neutralization realized in the available interfacial contact time (generally on the order of a few seconds in all types of scrubbers), depends largely on the rate of interfacial mass-transfer. The mass transfer rate, in turn depends largely on the physical and chemical form of the acid gases and the alkaline (basic) scrubbing reagents, and the physical characteristics of the systems used to bring them into contact with each other. The following sections outline the three most common approaches to alkaline scrubbing of acid gases."[1]

Wet Scrubbing

Wet scrubbing is an *absorption* process, in which the gas-phase is dissolved in a liquid a result of solubility. The dissolution rate is enhanced by using a mass-transfer device such as a packed-tower and a reactive scrubbing solution. The scrubbing solution is fortified to maintain the driving force for *absorption*. Both reactants are in solution as ionic species and only a small stoichiometric excess of material is needed to achieve high acid removal efficiencies. The rate-limiting step in wet scrubbing is the *absorption* of the acid gas into the liquid solution. The designs of mass transfer devices for this purpose focus on facilitating this *absorbtion* by counter-current gas/liquid flows, by maximizing interfacial area (between gas and liquid films), and by maintaining turbulent liquid and gas flow regimes. With the exception of large FGD (flue-gas desulfurization) systems on utility boilers, which employ limestone or hydrated lime slurries, the alkaline materials used in wet scrubbing are generally sodium based, since sodium salts are highly soluble in water. The most commonly used sodium based materials for wet scrubbing and their overall reactions with SO$_2$ are:

sodium hydroxide: $\quad$ $2NaOH + SO_2 \rightarrow Na_2SO_3 + H_2O$
sodium carbonate: $\quad$ $Na_2CO_3 + SO_2 \rightarrow Na_2SO_3 + CO_2$

The gas exiting a wet scrubbing system is generally saturated with moisture from the scrubbing liquid. High inlet loadings of insoluble particulate matter can be problematic for such systems.

Salts of acid gases removed by wet scrubbing systems are purged in scrubber blow-down and generally disposed as dissolved solids in wastewater. Depending on the process being controlled, this wastewater may also contain significant levels of insoluble, suspended solids.

Semi-dry Scrubbing

In semi-dry scrubbing the gas phase acid is also *absorbed* into a scrubbing liquid. However, rather than being a solution of alkaline material, it is typically a slurry prepared immediately prior to use. The reaction conditions are multi-phase and not as ideal as in a wet scrubber. A spray-dryer is typically employed as the contact device to avoid the problems of suspended solids in the liquid phase that occur in a packed tower. Interfacial area within the spray-dryer is maximized by mechanically atomizing the alkaline slurry into an open (usually vertical down-flowing) vessel in which gas flows are maintained in the turbulent regime, to facilitate mass transfer. Simultaneous heat and mass transfer from the hot gases to the slurry liquid transforms the aerosol particles of slurry into solid particles of acid salts and excess alkaline material that are collected downstream by a particulate capture device. Evaporation of the free water in the slurry results in quenching of the gases preferably at high temperatures. Salts of acid gases removed by semi-dry scrubbing systems are purged as solids with other captured particulate matter via an ESP or baghouse.

Dry Scrubbing

Acid gas removal by dry scrubbing also requires multi-phase mass transfer, but it occurs between the gas phase and the solid phase, rather than between the gas phase and the liquid and solid phases, as with wet and semi-dry scrubbing. The acid gas must first contact the solid surface of the dry alkaline material rather than being *absorbed* (dissolved) in a scrubbing liquid or slurry. Once the acid gas contacts the alkaline surface it is neutralized to form a solid salt that can remain as part of the solid particle alkaline material or "popped off" as a result of gas evolution. Chemically, these are less ideal neutralization reaction conditions than in wet or semi-dry scrubbers, since one reactant is in the gas phase and the other is in the solid phase. As a result, dry scrubbers usually require the greatest stoichiometric excess of alkaline material to achieve high acid removal efficiencies.

The dry sorbent injection (DSI) contact device can be simply the gas duct between the source and the downstream particulate capture device or a mixing chamber designed specifically to achieve contact time and mixing of the reactants. The key system performance factors are turbulent mixing of alkaline powder with hot gas while maximizing the time that the alkaline powder is suspended in the presence of the hot acid gases. There are systems in operation where DSI has been employed in conjunction with both a hot side and/or cold side ESP for particulate capture. Most studies have been done utilizing a fabric filter but commercially more sodium DSI systems are with ESP's. With an ESP the available acid-base contact time is limited to the time the alkaline powder and gas are in contact between the point of injection and the fields of the ESP. In a dry sorbent and semi-dry injection/fabric filter unconverted alkaline material in the suspended particulate matter becomes part of the filter cake on the bag filters and provides additional acid-base contact time as the cake accumulates on the bags between cleaning cycles. In some cases the sorbent can decrease the resistance of the filter cake improving the APC operation.

Unlike wet and semi-dry scrubbing, evaporation of free water in scrubbing liquor and quenching of gases is not inherent in the dry scrubbing process. Sorbent injection temperatures are a function of the process, sorbent and APC device.

Dry furnace injection (DFI) where the dry alkaline material is injected directly into the process furnace at high temperature has been tried in many in dustries.. The advantages of furnace injection are greater overall reaction time and higher reaction temperatures, producing faster reaction rates though this can upset the chemistry of the boiler creating other process concerns. Also, in certain the acid gases are generated by other APC devices. One such occurrence is SO$_3$ by the oxidation of SO$_2$ during NO$_x$ abatement.

TECHNOLOGY COMPARISON

"Each of the technologies described in the previous sections is capable of removing high percentages of both SO$_x$. "Hence, the choice of which technology to apply is usually driven by a combination of cost and ease of operation & maintenance. The primary operating objective is to maintain conformance with the applicable regulatory limits for acid gas release, while incurring the minimum annualized total cost. The annualized total cost comprises both capital and operating & maintenance (O&M) costs. The annualized capital cost is the repayment of principal and interest on the funds borrowed to purchase and construct the acid gas scrubbing system. The largest component of the annualized O&M cost is typically the cost of the alkaline reagent, but it also includes energy costs associated with fans and pumps, and the costs of treatment and disposal of liquid and solid residues. The following Table 1 summarizes and qualitatively compares the various characteristics of the different acid gas scrubbing technologies.

Table 1 – Acid Gas Scrubbing Technology Comparison									
TECHNOLOGY		**CHARACTERISTICS**							
		Capital Cost to Install	Efficiency (reagent utilization)	Neutralizing Reagent Cost	Energy Cost	Process/ Maintenance Complexity	Ease of Retrofit	Waste Water	Visible Plume
Wet Scrubbing	Calcium-based	HIGH	MEDIUM	LOW	HIGH	HIGH	LOW	YES	YES
	Sodium-based	HIGH	HIGH	HIGH	HIGH	HIGH	LOW	YES	YES
Semi-Dry Scrubbing	Calcium-based	HIGH	MEDIUM	MEDIUM	HIGH	HIGH	LOW	NO	NO
Dry Scrubbing	Calcium-based	LOW-MED	LOW-MED	MEDIUM	LOW	MEDIUM	MEDIUM	NO	NO
	Sodium-based	LOW	LOW-MED	MED-HIGH	LOW	LOW	HIGH	NO	NO

These advantages and disadvantages and their associated economics, form the basis for selection of an acid gas scrubbing system. Each type of system is best suited to a particular set of process and economic circumstances. The following paragraphs describe the dry injection technology in greater detail and outline the circumstances under which it is likely to be the best process technology choice."[2]

DRY SCRUBBING

When compared with wet scrubbing or semi-dry scrubbing, the primary disadvantage of dry scrubbing or dry sorbent injection is poor sorbent utilization. Wet scrubbing and spray drying mixes the acidic and basic species together to form a solution or slurry. The main factor that limits conversion and necessitates a slight stoichiometric excess of alkaline reagent, is the need for the continuous phase contactor (eg. packed column) is to provide a driving force for the reaction. A greater stoichiometric excess of alkalinity is required for a semidry scrubber than with a wet scrubber. Dry scrubbing has to contend with particle size disparity, the inherent disadvantages of heterogeneous gas-solid resistance, interference of both the reaction products (shrinking core theory) and the unreacted alkaline material thus resulting in the greatest requirement for stoichiometric excess of sorbent.

Despite these disadvantages DSI systems are extremely simple to install and operate and are very cost effective compared with the more capital intensive wet systems. The only rotating equipment required is a small blower for dilute phase pneumatic conveying of the dry alkaline sorbent to the process gas ducting and possibly a mill. "The following are examples of circumstances in which a dry scrubber is better suited to an application than either a wet scrubber or a semi-dry scrubber:

♦ The annualized cost of a dry injection system is lowest, or close enough to the other options to be influenced by factors to which costs cannot easily be assigned.

♦ The process throughput is not large enough to make semi-dry scrubbing economical (gas flows are less than approximately 100,000 acfm).

♦ Wastewater discharges from a wet scrubber cannot be permitted or are uneconomical to treat.

♦ Local electric power rates are high, making the higher system pressure drops associated with wet and semi-dry scrubbers too costly to operate.

♦ Moisture saturated exhaust gas or a visible steam plume from the stack are unacceptable.

♦ The process operation and maintenance complexity of wet and semi-dry scrubbers are incompatible with the facility operator's organization.

♦ Handling hazardous chemicals such as sodium hydroxide, calcium oxide, and calcium hydroxide are not acceptable.

♦ Particulate capture equipment (eg. ESP or fabric filter) is already in operation and acid gas controls need to be retrofitted."[3]

The remaining cost factor is the sorbent choice. That is:

1. The delivered cost of the sorbent per unit equivalent pollutant weight unit removed

2. The cost of spent sorbent disposal per unit pollutant removed.

Dry Injection Chemistry

Dry injection (DI) systems can employ either calcium or sodium alkaline materials for neutralizing acid gases. With a calcium based system, dry hydrated lime (purchased and stored as $Ca(OH)_2$) is directly injected and the overall neutralization reactions are essentially the same as those for the semi-dry scrubber:

$$Ca(OH)_2 + SO_2 \rightarrow CaSO_3 + H_2O$$

Sodium sorbent based systems the neutralization reactions are via sodium carbonate:

$$Na_2CO_3 + SO_2 \text{ in air} \rightarrow Na_2SO_4 + CO_2$$

The source of the carbonate can be synthetic sodium bicarbonate, naturally occurring sodium bicarbonate (nacholite), or naturally occurring sodium sesquicarbonate (trona). When these materials are injected into a hot (oxidizing) gas stream they "calcine" according to:

$$2NaHCO_3 (s) + O_2 + heat \rightarrow Na_2CO_3(s) + H_2O(g) + 2CO_2(g)$$

$$2(Na_2CO_3 \cdot NaHCO_3 \cdot 2H_2O) (s) + heat \rightarrow 3Na_2CO_3(s) + 5H_2O(g) + CO_2(g)$$

"The theoretical stoichiometry and mass requirements for the different alkaline materials and reactions are summarized in Table 2."[4]

Common Name	Chemical Name	MW	Neutralization Reactions	Theoretical Requirement		Purity	Practical Limit
				Moles base Mole acid	lbs base lb acid		lb base lb acid
Baking soda	Sodium bicarbonate	84.01	$2NaHCO_3 + SO_2 \rightarrow$ $Na_2SO_3 + H_2O + 2CO_2$	2	2.62	99%	2.65
			$NaHCO_3 + HCl \rightarrow$ $NaCl + H_2O + CO_2$	1	2.30		2.33
Hydrated lime	Calcium hydroxide	74.09	$Ca(OH)_2 + SO_2 \rightarrow$ $CaSO_3 + H_2O$	1	1.16	98%	1.18
			$Ca(OH)_2 + 2HCl \rightarrow$ $CaCl_2 + 2H_2O$	½	1.02		1.04
Nacholite	Sodium bicarbonate	84.01	$2NaHCO_3 + SO_2 \rightarrow$ $Na_2SO_3 + H_2O + 2CO_2$	2	2.62	97%	2.70
			$NaHCO_3 + HCl \rightarrow$ $NaCl + H_2O + CO_2$	1	2.30		2.37
Soda ash	Sodium carbonate	105.99	$Na_2CO_3 + SO_2 \rightarrow$ $Na_2SO_3 + CO_2$	1	1.65	97%	1.71
			$Na_2CO_3 + 2HCl \rightarrow$ $2NaCl + H_2O + CO_2$	½	1.45		1.50
Trona	Sodium sesqui-Carbonate	226.03	$2(Na_2CO_3 \cdot NaHCO_3 \cdot 2H_2O) + 3SO_2 \rightarrow$ $3Na_2SO_3 + 5H_2O + 4CO_2$	2/3	2.35	97.5%	2.41
			$Na_2CO_3 \cdot NaHCO_3 \cdot 2H_2O + 3HCl \rightarrow$ $3NaCl + 4H_2O + 2CO_2$	1/3	2.01		2.12

Table 2 – Theoretical Chemical Comparison of Dry Alkaline Sorbent Materials

Only the reactions between SO_2 and HCl and the alkaline material are shown in Table 2 for comparison purposes, but the stochiometry for SO_3/H_2SO_4 and other HX is essentially the same.

Although it is not used directly as a DI alkaline material, for reasons explained later, soda ash is also included for comparison. As can be seen from this table, calcium sorbents have an theoretical stoichiometric advantage relative to sodium sorbents, due to the divalent cation (Ca^{++}), as compared with the monovalent cation (Na^+). Even with differing molecular weights factored in, the approximate 2 to 1 advantage translates roughly into mass so that the theoretical mass requirement for the sodium sorbents is roughly twice the theoretical calcium sorbent requirement. However, the theoretical requirements do not account for differing reactivities and assume that equilibrium conversions are reached. The different alkaline materials used in dry injection exhibit different reactivities and due to the nature of the processes, equilibrium conversions are never reached. A common measure of actual alkaline sorbent usage as compared with the theoretical requirement is the nomalized stoichiometric ratio or NSR, defined as:

$$\text{normalized stoichiometric ratio} \equiv \frac{\text{moles alkaline material actually injected}}{\text{moles alkaline material theorectically required}}$$

The well established axiom that combustion reactions are favored by "time, temperature, and turbulence" can also be applied to dry acid gas scrubbing. The process factors that have the most effect on alkaline reactivity and conversion are:

♦ available interfacial surface area,

♦ reaction temperature, and

♦ reaction time

♦ Turbulence/mixing

Surface Area

In dry scrubbers available area is determined by how much powdered alkaline material is suspended in the hot gas, by the individual alkaline material particle size, and by the average available surface area of each particle. Up to the point where solid particles of alkaline material drop out of the gas stream due to insufficient velocity to keep them suspended, the amount of alkaline material in the gas is driven by inlet acid gas loading. Individual particle size for "sorbent grade" alkaline materials is generally minimized by mechanical refinement to a practical mean diameter of 20 to 30 microns. In dust injection of sorbent being milled on site can improve utilization by lowering the particle size below this number. Another factor that has a great potential to effect available surface area for reaction is the interstitial pore area of the particles. Sodium sorbents such as sodium bicarbonate and trona are porous and their internal surface area can be several orders of magnitude greater than the outer surface area of the particle.

Although this characteristic is of limited importance with hydrated lime, it is significant with sodium sorbents. The reactions shown in Equations 12 and 13 above result in the explosive evolution of $H_2O(g)$ and $CO_2(g)$ from the source carbonate molecules, causing the solid particles to "pop-corn", greatly increasing the available surface area of sodium carbonate within the particles of alkaline sorbent material. Sodium carbonate (soda ash) itself is rarely used in dry scrubbing because this calcining does not occur and the surface area available for neutralization reaction, even with finely milled powder, is much less than in the "pop-corned" particles formed by calcining sodium bicarbonate/sesquicarbonate in the hot process gases.

Reaction Temperature

Although acid-base neutralization reactions are exothermic and the theoretical reactant conversion is negatively effected by temperature, as a practical matter the positive exponential temperature dependence of the reaction rate factor is far more important relative to conversion of alkaline material. Up to a point, the higher the gas temperature the faster the neutralization reaction and the greater the conversion of alkaline material to acid salts in the available reaction time. Dry injection systems often operate in the range of 250°F to 1500°F. Trials at many different process show that injection of trona 675°F can significantly increases the reactivity and conversion of the alkaline material, greatly reducing the effective NSR.

Reaction Time

"One of the aspects of dry scrubbing that can significantly offset the inherent limitation on reactant conversion is reaction time. In wet scrubbers the available reaction time is the time that the gas stream and scrubbing liquor are actually in contact. This time is generally on the order of a few seconds. In SD and DI systems that use ESPs for particulate collection, the situation is analogous. The available reaction time is the time that the gas stream and alkaline material are in contact - the time between injection of the alkaline material into the hot gas and the ESP fields removing the suspended particulate material.

In DSI/FF (and SD/FF) systems the neutralization reaction is a 2-phase process. Much of the conversion of alkaline material to acid salts occurs while the alkaline material is suspended in the gas but significant additional conversion can occur in the baghouse filter cakes between filter bag cleaning cycles. The typical contact time between reactants while the sorbent material is suspended in the gas stream is on the order of a few seconds. Exposure of alkaline material in the filter cakes to acid gas can range from several minutes to hours, depending on the baghouse cleaning cycle and the actual system pressure drop due to inlet dust loading. This additional residence time can increase the alkaline material conversion from a "once through" value of 30 to 50 percent, up to between 60 and 80 percent.

In calcium-based DI systems, the alkaline particle surface area does not benefit from the "pop-corn" effect of sodium materials. As a means of increasing sorbent utilization, calcium based DI processes often include systems for recycling collected particulate matter containing unreacted calcium in order to increase the conversion of alkaline material. Baghouse bottoms are collected and reinjected into the hot gas stream, increasing the total alkaline residence time and thereby the overall alkaline utilization. Depending on the system, single-pass conversions can range from 30 to 60 percent. Systems equipped with recycle can exceed 70 percent conversion of calcium. The disadvantage of sorbent recycle is that the dust loading in the hot gas going into the particulate capture device is significantly higher. With existing ESPs this may not be feasible and the particulate capture performance of existing baghouses may be significantly degraded. New baghouses designed for that dust loading will be much larger than they would be with the once-through dust loading and the greater the inlet dust loading the more frequently the bags will be cleaned. With sodium-based DI systems, the higher particle surface area available enables more of the alkaline reactant conversion to occur while the material is initially suspended in the hot gas stream and allows greater conversion in the filter cakes on the bags in the baghouse. Hence, sodium-based systems are able to achieve high alkaline conversions without recycling baghouse bottoms to the hot gas stream, reducing the inlet loading of particulate matter that includes a significant fraction of material that is essentially inert, and thereby the required sizing of the particulate capture device"[5].

Turbulence

The effect of mixing/turbulence has become increasing important to achieve high utilization efficiencies of sorbents. It has been determined that the sorbent as well as the acid gases can channel or profile as the acids and sorbent travel down the duct to the particle control device. Mixing devices down stream from the injection point prior to the particulant collection device is important to improve the utilization effiency of the trona especially when mitigating SO$_x$. Computer modeling should be considered to obtain the most cost effective DSI system.

Trona as a Sorbent for DSI

Recently several coal fired power plants have begun using dry sorbent injection of T200 a mechanically refined trona product. Dry sorbent injection is preferable to liquid injections due to the ease of handling, reduced risk of fouling downstream duct or equipment, and improved safety (including lower conveying pressures). Since T200 is a sodium compound it conditions the ash to lowers its resistivity having a positive effect on the operation of both cold and hot side ESP's. Test results using the controlled condensation method verified visual indications that stack SO$_3$ levels below 10 ppm were routinely attainable without opacity excursions.

"Trona is a sodium based sorbent that is relatively low in cost, is highly reactive, and requires minimal capital for injection into the duct. Trona is typically injected in flue gas streams in various processes for acid gas mitigation. The use of trona for SO$_x$ mitigation is well established. The use of sodium conditioning as an alternative to SO$_3$ conditioning has been documented."[6] The practical application of dry sodium injection for simultaneous SO$_3$ mitigation and ash conditioning is the subject of this report.

Background of Trona Ore

Trona's formation at Green River, Wyoming is the result of rapidly repeated evaporation cycles of Lake Gosiute approximately 50-60 million years ago. The original lake was fresh water and supported abundant flora and fauna. When the climate changed from humid to arid, Lake Gosiute evaporated and trapped the remnants of the once abundant life. The lake bottom became a mixture of mud and organic sediments that formed oil shale. Runoff water from the nearby mountains continued to supply sodium, alkaline earth, and bicarbonate to the lake. Since the rate of evaporation was high, the clear waters changed to brine that finally precipitated a sodium carbonate-bicarbonate compound known as the mineral "trona" (chemical formula: Na$_2$CO$_3\bullet$ NaHCO$\bullet$2H$_2$O). Alternating climates prevailed for about two million years. Periods of rains washed mud into the lake to cover previously formed carbonate type precipitates while interim periods of arid climate produced new precipitates. This caused numerous beds of trona to be formed. In total, they contain billions of tons of trona and related minerals.

The tropical rains eventually returned to expand Lake Gosiute. This washed sand and mud from the surrounding mountains into its waters. Sediments of clay and shale built new formations as the geological evolution of the Green River Basin progressed. Buried under these sediments remain the trona beds that have become the basis of the present soda ash and sodium-based industry located around Green River, Wyoming.

Solvay Chemicals, Inc. currently mines trona ore from bed number 17 located at an approximate depth of 1500 feet. This ore body under Solvay's lease holdings is approximately 12 feet thick and of very high quality. Bed 17 will provide ore for many years; however, should the need arise, additional trona ore may be extracted from other beds as well.

Solvay Chemicals employs the "room and pillar" method of mining. Long drifts (tunnels) are mined with cross cuts at specified distances and the remaining pillars support the roof of the mine.

Solvay currently uses boring type miners and long wall mining to extract the ore. Roof bolting machines insert steel rods into the roof of the mined areas. This supports the immediate rock roof and prevents rock falls. Once cut, the ore is transported via shuttle car to a series of belt conveyors which discharge into underground storage bins. From these storage bins, the ore is hoisted to the surface in 20-ton capacity skips.

Solvay offers mechanically refined sodium sesquicarbonate (natural trona) to the merchant market. Coarse product exclusively is marketed and sold to the animal feed market by an outside company. Alternative products, Solvay T –50 (250-300μms) and T- 200 (23μms), are marketed directly by Solvay for acid gas and acid neutralization applications.

Trona product operations are located in the ore-crushing portion of the soda ash plant. Trona ore is crushed and screened before being sent through a fluid bed dryer. The dryer serves to remove free moisture from the product and to separate the coarse (T-50) from fine (T-200) product. Trona products are stored in bulk quantity awaiting rail and truckload out. A bagging operation exists for the packaging of T50 products into 50-pound bags and both T50 and T200 in 2000-pound bulk sacks.

De-acidification of Acid Gases with Trona

Trona is rapidly calcined to sodium carbonate when heated at or above 408°K (275°F). The "popcorn-like" decomposition creates a large and reactive surface by bringing un-reacted sodium carbonate to the particle surface for acid neutralization. The by-products of the reactions are sodium salts. For example the SO$_2$ chemistry is:

1. $2(Na_2CO_3 + NaHCO_3 + 2H_2O) + 3SO_2 \rightarrow 3Na_2SO_3 + 4CO_2 + 5H_2O$
2. $3Na_2SO_3 + 1.5O_2 \rightarrow 3Na_2SO_4$

Sodium carbonate is brought in contact with acid vapor in the flue gas stream by a simple direct injection of a sodium compound such as trona that will thermally decompose to form sodium carbonate. After the initial sorbent surface of the sodium carbonate has reacted with SO$_2$ to form sodium sulfite, or sulfate, the reaction slows due to pore blockage -resisting gas phase diffusion of SO$_2$. In order for the reaction to continue, the sorbent particle must decompose further. This decomposition evolves H$_2$O and CO$_2$ gases into the surrounding atmosphere creating a network of void spaces throughout the particle. This process exposes fresh reactive sorbent and allows

SO_2 once again to diffuse into the particle interior. **This** increase **in** surface area ('popcorn effect") i s on the order or **5** to 20 times the original surface area."

The particle size of the **trona** sorbent can have a dramatic effect on the removal efficiency of SO_2 as can be seen in Figure 1 :

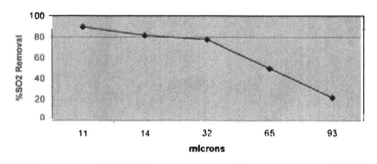

Figure 1. Effect of Sorbent Size @ NSR 1

COLD SIDE ESP CONDITIONING'

"Referring to Figure **4,** at a DSI rate of 2.9 TPH with the **SCR** O/S there is an indication that ESP conditioning may improve SO_3 removal efficiencies. The first data p i n t was taken directly after trona injection began. There was no attempt to condition the ESP prior to this first test point. This resulted in a removal efficiency of about 63%. The removal was measured again at 2.9 TPH later in the test schedule. However, for the second data point at 2.9 TPH the ESP was well conditioned and a **78%** removal was seen. Looking at these two data points in Figure 2 one can see that the unconditioned point is slightly above a 2.0 molar ratio (removing 63% SO_3) while the conditioned p i n t is actually removing **a** higher percentage of SO_3 **(78%)** while being at a molar ratio slightly lower than 2.0.

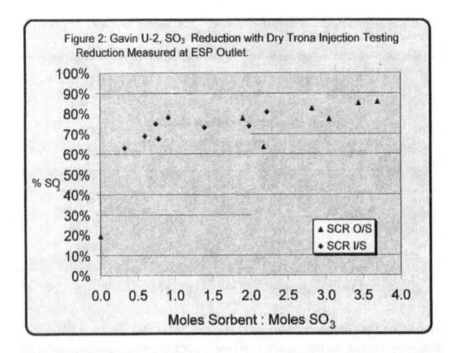

Figure 2: Gavin U-2, SO$_3$ Reduction with Dry Trona Injection Testing Reduction Measured at ESP Outlet.

Hydrated lime injection rates were limited to 2 to 2.5 TPH due to ESP performance degradation. No such degradation was noted with trona injection, see Figure 5. It did not appear that ESP performance was negatively impacted by trona injection. VI curves were also generated during trona injection with no indication of back corona. After a permanent system was installed, it was noticed that plant operators actually injected trona for the sole purpose of enhancing ESP performance. This was done when several T/R sets were O/S due to internal grounds. Hence, the trona system actually benefited ESP performance while SCRs were O/S."[6]

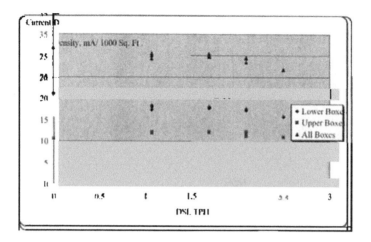

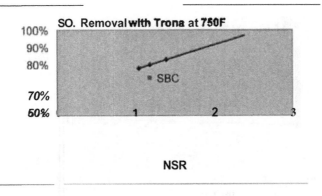

Figure 4

The data shown in the graph below is from the same commercial process in which mechanically refined trona was injected into the flue gas stream at a 1.5 normalized stoichiometric ratio (NSR} rate while varying the injection temperature. Removals as high as 91% were achieved injecting T200 into a hot ESP w/o a baghouse. At this writing the T200 injection is being done at 725°F using 25% less trona than sodium bicarbonate and getting the same SO, removal. SO, removals as high as 91% have been demonstrated using trona at 725% and at an NSR of 1.5. Please see below:

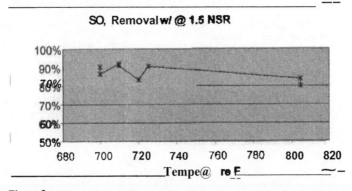

Figure 5

As can be seen in the graphs above high SO, removals can be achieved at higher temperatures than has been typically observed at lower temperatures unless a baghouse was used. Also. lower NSR values are needed for the high SO, removal than observed in laboratory studies or actual plant operating data than milled sodium bicarbonate unless a baghouse was used.

Prior to injecting trona the "perf plates" of the hot side ESP would plug within days of injecting sodium bicarbonate but when using trona no pluggage was encountered and the system is operating trouble frre for the last year.

CONCLUSIONS

In certain acid gas control applications. DSI is the simplest and least costly process to implement (new or retrofit), and the least costly to operate and maintain. Both calcium and sodium alkaline reagents are well proven in DSI applications. However, calcium hydroxide is a hazardous chemical with potential worker exposure hazards, whereas sodium bicarbonate/sesquicarbonate are considered to be nuisance dusts from the occupational exposure standpoint. The delivered cost of calcium and sodium reagents varies with location, but is more consistent with the chemically processed materials, dry hydrated lime and sodium bicarbonate, than with the naturally occurring materials nacholite and trona, which only have natural deposits in certain parts of the country. The sodium sorbents are generally more costly than the calcium sorbents, although the material cost (less freight) of dry hydrated lime and trona are similar. Among the sodium sorbents, nacholite and synthetic sodium bicarbonate are slightly more efficient for SO$_x$ reduction, while trona is slightly more effective for HX reduction. The delivered cost of synthetic sodium bicarbonate is significantly higher than nacholite, and nacholite is generally higher than trona.

REFERENCES:

1. Comparison of Dry Injection Acid-Gas Control Technologies; John Maziuk

  Solvay Minerals, Inc., P.O. Box 27328, Houston. TX USA (john.maziuk@solvay.com)

  John H. Kumm EA Engineering, Science & Technology, 15 Loveton Circle. Sparks, MD USA (jkumm@eaest.com)

2. Ibid

3. Ibid

4. Ibid

5. Ibid

6. Successful Mitigation of SO$_3$ Emissions While Simultaneously Enhancing ESP Operation at the General James M. Gavin Plant in Cheshire, Ohio by Employing Dry Sorbent Injection of Trona Upstream of the ESP, Ritzenthaler and Maziuk, Sept. 2004.

7. *Ibid*

8. Successful Mitigation of SO$_3$ by Employing Dry Sorbent Injection of Trona Upstream of the ESP, John Maziuk, November 2005.

9. Successful Mitigation of SO$_3$ Emissions While Simultaneously Enhancing ESP Operation at the General James M. Gavin Plant in Cheshire, Ohio by Employing Dry Sorbent Injection of Trona Upstream of the ESP, Ritzenthaler and Maziuk. p2. Sept. 2004.

ENERGY

ENERGY

ENERGY BALANCES OF GLASS FURNACES: PARAMETERS DETERMINING ENERGY CONSUMPTION OF GLASS MELT PROCESSES

Ruud Beerkens
TNO Science & Industry
Eindhoven, The Netherlands
ruud.beerkens@tno.nl

ABSTRACT

Energy balance models have been developed and are tested for different types of industrial glass melting furnaces: recuperative, regenerative and oxygen fired furnaces. Modelling studies show the quantitative effects of batch humidity, cullet % in batch, combustion air or oxygen excess, flame emission coefficient, pull, batch and/or cullet preheating, temperature control and some other process or design parameters on the fuel / electricity consumption. Results for regenerative furnaces show that batch humidity and air excess should be controlled to avoid excessive energy consumption. The energy consumption and the energy balance of one of the most efficient glass furnace types, found in a benchmark investigation, are presented.

INTRODUCTION

Annual averaged energy consumption figures of glass furnaces in the container, glass fibre and float glass sectors have derived from production and operation data of a large number of furnaces (in total about 280 furnaces) for the years 1999 [1] and 2003 by TNO [2].

From these data, the specific energy consumption (energy demand per ton molten glass) values for each investigated furnace have been derived and these values are expressed in primary energy equivalent. The energy supply to the melting furnace (excluding energy consumption of separate working ends or feeders and excluding fans) only: such as fuel, electricity and oxygen is taken into account, but calculated as primary energy demand. Electricity is used for oxygen separation from air by mol-sieve absorption or cryogenic techniques and for electric boosting electrodes. In the calculations of the equivalent specific primary energy consumption (GJ primary energy per ton molten glass), we assume:

- an electric power plant efficiency of 40 % (1 kWh requires 9 MJ fossil combustion energy-net calorific value) and
- 0.375-0.4 kWh electricity for 1 m_n^3 pure oxygen generation (1 m_n^3 = 1 m^3 gas at 273.15 K and 101.3 kPa). Thus the primary energy equivalent for pure oxygen is about 3.4 to 3.6 MJ per m_n^3 oxygen.

The conversion of GJ/metric ton based on net calorific combustion value of the fuel to gross value in MMBTU/short ton: 1 GJ/metric ton = 0.95 gross MMBTU/short ton.

The net calorific value for fossil fuel combustion is used in this benchmark study. The gross combustion value is about 10 % higher than the net calorific value of natural gas firing.
The benchmark procedure has been described previously [1-3] and is used to identify the most energy efficient furnaces in the different glass industry sectors, such as in the container glass industry.

Generally, some normalisation of the energy consumption figures is applied to take into account the cullet % in batch or the age of the furnace. Statistically found dependencies of observed energy consumption levels versus cullet % in the batch or versus furnace age are used to apply the normalisation rules/equations [2].

In the next section, the main observations and conclusions of the benchmark studies are summarized.

The glass industry is confronted with excessive increases in fuel costs in recent years plus the introduction of CO_2 emission trading in the European Union [4] since 2005, therefore it is of great importance to find ways to optimize energy efficiency in glass melting operations.
Section 3 shortly explains the derivation of energy balance models for glass furnaces and the use of such models to investigate the effects of process conditions on energy consumption of industrial glass furnaces. These models are applied and tested for existing industrial glass furnaces. Results of the model calculations are shown for typical float and container glass furnaces.

Section 4 gives a summary of possible measures that can be applied to save energy and the last sections give general conclusions.

ENERGY EFFICIENCY BENCHMARKING OF GLASS FURNACES

Energy benchmark investigations for glass-melting tanks have been performed by TNO for the full production years 1999 en 2003 to fulfil the request of the Dutch government to carry out these energy efficiency benchmark studies every 4 years (Voluntary Agreement on Energy Efficiency Benchmarking and striving to obtain efficiencies of all glass furnaces in the Netherlands within the top 10 % (worldwide benchmark)). Results of the 1999 benchmark study have been published elsewhere [1]. In this paper, the main observations of the benchmark studies for these years are summarized.

Figure 1 shows the ranking of the specific primary energy equivalent consumption (GJ/metric ton molten glass) for individual glass furnaces from lowest to highest figure for the container glass industry in 1999 and 2003.

This figure shows almost similar profiles of the curves for 1999 and 2003. For the production year 1999, 131 glass furnaces and in 2003, 90 container furnaces have been investigated.

Information of furnaces from all over the world, but most data from Europe are included in the benchmark study, however the number of investigated furnaces is only a fraction (< 15 %) of the total number of existing container glass furnace installations world-wide.

The energy consumption values presented in figure 1 are normalized to primary energy equivalent (taking into account the primary energy demand for electricity and oxygen production) and 50 % cullet in the batch (50 % of the glass is molten from recycled glass cullet).

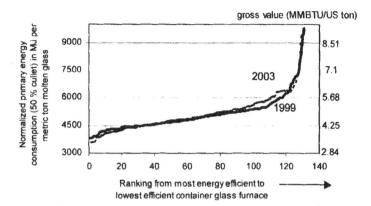

Figure 1   Ranking of the specific energy consumption values of container glass furnaces in 1999 & 2003. Values normalized to primary energy equivalent and 50 % cullet. The furnaces investigated in 1999 are partly different from the set of furnaces of 2003. Energy demand refers only to the melting tank (excluding feeders, & working-end). The calorific energy consumption per short ton (US) is given on the right vertical-axis based on gross calorific combustion value of the fuel.

The energy consumption of the most energy efficient furnaces decreased when comparing the 2003 to the 1999 benchmark, see figure 1 (from about 3850 down to 3600 MJ/metric ton molten glass or from 3.3 down to 3.1 net MMBTU/short ton, this is about 3.65 down to 3.4 gross MMBTU/short ton), but the average value (about 5100-5150 MJ/metric ton molten glass = about 4.825 gross MMBTU/short ton, not normalized to fixed furnace age) remained the same. It should be mentioned that the set of furnaces of the 1999 benchmark is not exactly the same as the set of furnaces in the 2003 inventory.

The following observations have been derived from the benchmarking results:

Container glass sector:
- The most energy efficient container glass furnace operating in 2003 consumes about 3.5-3.6 GJ primary energy per metric ton molten glass (based on 50 % cullet in batch) or 3.3 to 3.4 gross MMBTU/short ton glass melt, this is an end-port fired regenerative furnace and the energy demand has been normalized to new state (age = 0);
- The average energy consumption values of the end-port fired furnaces in the 1999 benchmark are almost similar to the average energy consumption of all oxygen-fired furnaces (taking into account the energy consumption for oxygen generation) and significantly lower than most cross-fired furnaces. In the 1999 benchmark no normalisation for furnace age has been applied.
- In the benchmark of 2003 data, the average specific energy consumption (4.3 GJ/ton molten glass) of a limited set oxygen- fired container glass furnaces is even lower than the average energy consumption of end-port fired regenerative furnaces (4.6 GJ/ton molten glass = 4.35 gross MMBTU/short ton glass melt), based on 50 % cullet, new furnace state

and taking into account the energy demand for oxygen generation. The average value of cross-fired regenerative furnaces in 2003 was only slightly higher: 4.7 GJ/ton molten glass (4.45 gross MMBTU/short ton). The set of recuperative container glass furnaces show much higher values (on average about 40 % higher energy consumption levels compared to the regenerative furnaces).

- For the 1999 investigations, 7 of the 10 most energy efficient furnaces appeared to be end-port regenerative furnaces and in 2003, 8 or 9 of the 10 most energy efficient furnaces are end-port fired. Cross-fired furnaces with batch preheating generally show also rather low energy consumption levels. But, several furnaces using batches with 50 % cullet operate at less than 4 GJ energy per metric ton molten glass (< 3.8 gross MMBTU/short ton glass melt), even without batch preheating.
- Melting of cullet, instead of batch, saves on average 29 % energy in the container glass sector;
- Although the energy consumption generally does not increase linear with time (age of furnace), an average increase of 1.35 % energy per year extra age has been found statistically.
- The band width of energy consumption figures for 80 % of the furnaces (10 % most and 10 % lowest energy-efficient furnaces not taken into account) is rather small in 1999: 80 % of the investigated furnace showed an energy consumption (primary energy equivalent based on 50 % cullet) between 4.3 and 6.25 GJ/metric ton glass melt (4.07-5.9 gross MBTU/short ton) and in 2003, 80 % of the furnaces operated between 4.2 and 6.35 GJ/metric ton molten glass.
- The 2003-benchmark results (90 furnaces) show after normalisation of the energy consumption data to 50 % cullet in the batch and new furnace conditions that the average container glass furnace consumed 4.72 GJ/metric ton molten glass (4.46 gross MMBTU/short ton glass melt). 10 % of these furnaces in the 2003 benchmark analysis consumed less than 3.81 GJ/metric ton molten glass (3.60 gross MMBTU/short ton glass melt).
- The glass melt pull rate or specific pull rate (tons glass per m^2 per day) generally has an significant influence on the specific energy consumption. However, for container glass furnaces operating at a pull rate above 250-300 metric tons per day or > 2.5-2.75 metric ton per m^2/day, there is hardly a clear trend of further decreasing specific energy consumption values as specific pulls increase, see also Trier [5] and Fleischmann et al [6];
- The average energy consumption of container glass furnaces hardly depends on the produced glass colour. The average energy consumption of flint glass furnaces is within about 1-2 % the same as for coloured glass melting.

The benchmark study shows that well-insulated end-port fired glass furnace with regenerator efficiencies above 60 % (more than 60 % of the sensible energy of the flue gas generated by combustion - reference temperature is 0 °C - is transferred to combustion air) can operate at specific energy consumption levels below 3.6 GJ per metric ton molten glass (3.40 gross MMBTU/short ton glass melt) when using 50 % cullet batches. But, some of these furnaces even show energy consumption values around 3.4 GJ/ton (3.2 gross MMBTU/short ton) for > 80 % cullet levels.

Oxygen-fired furnaces with cullet/bath preheating also showed values close to this range [7, 8], including the energy demand for oxygen production.

Figure 2 shows the statistics of the specific energy consumption levels normalized to primary energy and 50 % cullet for the different furnace types applied in the container glass industry.

This figure shows that regenerative and oxygen-fired furnaces show about the same distributions of energy consumption levels. Typically 50 % of all regenerative and oxy-fired furnaces investigated in the benchmark analysis operate at levels less than 4.5 GJ/metric ton molten glass (50 % cullet, total primary energy equivalent, furnace at new conditions) = 4.25 gross MMBTU/short ton glass melt.

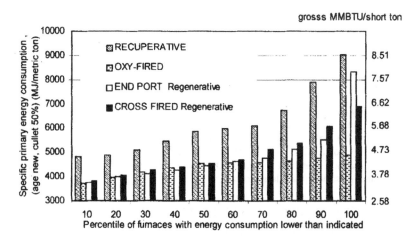

Figure 2. Energy consumption distributions for different types of container glass furnaces., Data are derived from the 2003 energy efficiency benchmark study for container glass furnaces. The graph shows on the vertical axis the maximum values of spe- cific energy consumption for a percentage of furnaces given on the horizontal axis. All values normalized to primary energy, 50 % cullet, including electricity & oxygen, new furnace state. Values on right vertical axis in gross MBTU/short ton

Float glass sector:
- The specific energy consumption is very strongly dependent on the size of the furnace. A furnace with a glass production of typically > 800 metric tons (molten glass) per day requires about 10-12 % less energy compared to a furnace producing about 500 metric tons float glass melt per day.
- The ageing of the furnace leads year on average to 1 to 1.3 % more energy consumption per;
- The effect of cullet in the batch on energy consumption could hardly been statistically investigated from the process data of the set of float glass furnaces, because of the small range of cullet ratios applied when comparing all these furnaces (typical cullet ratios in float glass batches: 20-30 %);
- The most efficient large float glass furnaces (> 600 metric tons per day) operate with 5.3- 5.8 GJ/metric ton molten float glass (5 – 5.5 gross MMBTU/short ton molten glass), based on 25 % cullet in the batch. This means that about 30-35 % of the energy supplied to the furnace is used to heat the glass plus reheating the re-circulating glass and for chemical reactions during batch melting. On average, the investigated float glass furnaces (from a dataset of about 40 furnaces) showed a specific energy consumption of 6.5 to 7 GJ/ metric

ton molten glass (6.15-6.6 gross MMBTU/short ton molten glass), normalized to 25 % cullet in the batch

The energy consumption levels of the investigated float glass furnaces ranges from 5.7 GJ/metric ton molten glass to 8.7 GJ/ton glass melt (from 5.4 up to 8.2 gross MMBTU/short ton molten glass). The energy consumption however depends very much on furnace age and pull rate in this industry.

Fiber glass industry (limited information available because of confidentiality reasons):
The most energy efficient furnaces appeared to be oxygen-fired furnaces in this sector, since no regenerative furnaces are used here, and recuperative furnaces demand on average about 25 % more energy in this sector than the oxygen-fired furnaces, taking into account the energy required to generate the oxygen.

The energy consumption levels including the energy consumption of the E-glass furnace refiners and forehearths and normalized to primary energy are between about 8 and 25 GJ/metric ton molten glass (between 7.5 and 23.5 gross MMBTU/short ton molten glass) and on average about 12 to 12.5 GJ/metric ton molten E-glass (± 11.5 gross MMBTU/short ton molten glass). E-glass furnaces generally produce less glass (50-120 tons glass per day) than container glass and float glass furnaces.

ENERGY BALANCES OF GLASS FURNACES

Generally energy is supplied to glass furnaces by:
- combustion of fuel ($Q_{combustion}$),
- preheated combustion air ($Q_{preheatair}$),
- electric power ($Q_{electric}$),
- sensible heat of fuels (Q_{fuel}), oxygen (Q_{oxygen}) or false air (Q_{air})
- (preheated) batch (Q_{batch})

This energy supply splits up in:
- glass melt heat capacity $Q_{heatmelt}$ (heat capacity of melt exiting furnace boundary),
- reaction enthalpy for decomposition and fusion of the raw materials, $Q_{reaction}$,
- evaporation energy for water $Q_{water\ evaporation}$,
- heat contents of the emitted flue gases, Q_{flue} (including batch gases),
- heat fluxes through the different walls, including the burner ports $Q_{structure}$,
- forced cooling losses (electrode cooling, stirrer cooling, cooled bar), Q_{cool},
- heat fluxes of combustion gas leakage through open joints and radiation heat loss through peepholes, $Q_{leakage}$ and $Q_{radiation}$ or burner ports,
- and the rate of the enthalpy increase of the system itself, $Q_{system\ heating}$.

In most energy balance models, the last term $Q_{system\ heating}$ is set on zero, assuming stationary conditions. However, especially in regenerative furnaces, cyclic variations (due to fire cycles) in the heat input will lead to fluctuations in the total heat contents of the system (furnace plus glass melt) and cause variations in air-preheat temperatures.

In the energy balance models, the reference state should be unambiguously defined: the heat contents of the mass flows of gas and molten/solid material should be based on fixed standard reference temperature, for instance, 273.15 K or 295 K. In this study the reference temperature is taken on 273.15 K.

The boundaries of the system on which the energy balance has to be determined should be unambiguously defined also. In most cases, the system boundary for the energy balance is fixed by the entrance of throat, the furnace walls, the flue gas exit of the regenerators or recuperator(s) and the charging end in the doghouse.

In the case, that relatively cold return glass melt flow from the refiner or working end to the melting end occurs, this extra energy transfer (extra hot melt flowing from melting-end into the throat, and colder melt flowing back from refiner into melting-end) should also be taken into account (reheating of colder re-circulating glass melt). Especially, in throats that are worn-out during the furnace campaign, re-circulation ratios (return flow versus net pull) may increase with time. In float glass furnaces, re-circulation of glass melt (extra energy supplied to reheat the return glass melt flow) from the working end will affect the energy balance of the melting end, this return flow may even cause more than 8-10 % extra energy demand. The re-circulating flow of melt from the colder working end back into the melting end may be as large or even larger than the net pull: Thus, relatively cold glass melt flows back from the working end to the melting end and consequently the hotter forward flow into the working end has to increase (= net pull plus return flow). The re-heating of the re-circulating melt is taken into account in the energy balance models. For float glass furnaces the energy flow of the exiting glass including the re-heating of re-circulating glass is taken into account in $Q_{heatmelt}$.

In detailed energy balances, radiation from the furnace through the burner ports to the regenerators can be taken into account in both the energy balance of the furnace itself and the energy balance of the regenerators. Also for these air preheating devices (regenerators or recuperators), energy balances may be used, taking into account the heat contents of the incoming flue gases and exiting flue gases, the incoming ambient air and exiting preheated air, the structural losses and leakage plus the incoming heat radiation from the furnace through the exhaust port.

Energy balances of glass furnaces require:
- measurements of (time-averaged) flue gas temperatures,
- calculation of the flue gas volume flows plus flue gas composition (from fuel consumption and excess air plus the batch gases (CO_2, water vapour, SO_2, etcetera),
- detailed information on the construction of the furnace (insulation layers, burner ports, non-insulated parts, cooling devices for electrodes/stirrers/sidewall) [5];
- enthalpy of the raw materials and gases (heat capacity) and
- heat capacity information of the exiting glass melt,
- plus the standard reaction enthalpy associated with the decomposition and fusion of the raw materials [9].

From the oxygen concentration in the flue gas, the fuel composition and fuel and oxygen consumption and the batch composition plus the batch-charging rate, the exhaust gas (batch gases + combustion gases + excess air or oxygen) volume flow and composition can be determined and from this information the flue gas heat contents per °C can be derived.

Table 1 (see also literature ref. 3)

Energy input		Variables needed to calculate/determine energy flow
Fuel	Q_{fuel}	heat capacity supplied fuel flow, fuel consumption rate
Fuel combustion	$Q_{combustion}$	combustion enthalpy (net calorific value), fuel consumption rate
Oxygen or false air	Q_{oxygen}	sensible heat of oxygen supplied to the combustion process
Preheated air	$Q_{primary\,air}$	air flow (from air demand and excess air), average heat capacity of air, preheat temperature
Electric boosting	$Q_{electric}$	power supply (net) to electrodes
Batch	Q_{batch}	batch heat capacity, batch charging rate, humidity of batch, temperature of batch
Recirculating glass melt enthalpy	$Q_{rec\,melt}$	recirculation ratio, average temperature of returning glass melt from refiner/working end
False or primary cold air	Q_{air}	sensible heat of false air or primary air entering the furnace
Energy output		**Variables needed to calculate/determine energy flow**
Glass melt enthalpy	$Q_{heat\,melt}$	average temperature, heat capacity and flow of glass melt into throat or neck to the working end
Evaporation enthalpy water	$Q_{water\,evaporation}$	water content batch, evaporation enthalpy
Wall heat losses	$Q_{structure}$	compilation of layers of refractory and insulation for each wall segment, thickness of each layer, heat transfer coefficient at outside (may be effect by forced cooling), heat conductivity of each layer
Flue gases	Q_{flue}	composition of flue gases, volume flow of flue gases (combustion gases + excess air + batch gases + evaporated water), average heat capacity for this composition, temperature at system boundary
Cooling Losses	$Q_{cooling}$	heat carried away by cooling air or water (volume flow x spec. heat capacity x temperature increase cooling medium)
Radiation losses	$Q_{radiation}$	surface area of radiating furnace parts to environment, emissivity of these surfaces and temperatures
Leakage losses	$Q_{leakage}$	volume flow out-leaking combustion gas & temperature

The specific heat of the gas components is averaged over the temperature range from reference temperature to exhaust temperature. For natural **gas** firing the average specific heat enthalpy $(\mathrm{J \cdot m_n^{-3} \cdot K^{-1}})$ is about 10 % higher compared to the c_p of the air (based on heat capacity per volume unit gas at reference pressure (=1 bar) and temperature **273.15** K). Per 1 m_n^3 combustion air, typically about **1.12 – 1.2** m_n^3 combustion gas is formed.

Since glass furnaces hardly operate fully stationary and cyclic processes, such as the fire cycle in regenerative furnaces take place, values averaged over at least 1 complete cycle (but it is better to take the average over much longer time periods) should be used in the overall (assumed as steady state) energy balance model.

In most cases, the energy losses by the flue gases, the enthalpy of melting and sensible heat of the exiting glass melt and the heat losses through the walls can be estimated fairly accurate, but the heat losses by diffuse cooling or leakage are often very **difficult** to access. These energy losses are often estimated by balancing the energy input and **known** energy losses/output, the rest being the unidentified energy losses. These diffuse heat losses may contribute from **5** to more than 10 % of the total energy input. Figure 3 shows the energy losses/energy output from a typical **float** glass furnace, producing 600 metric tons molten glass/day (= 661 US tons/day).

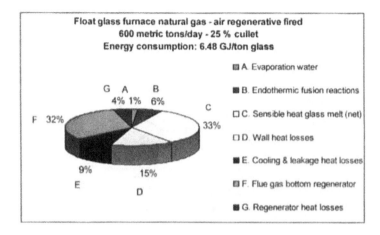

Float glass furnace natural gas - air regenerative fired
600 metric tons/day - 25 % cullet
Energy consumption: 6.48 GJ/ton glass

- A. Evaporation water
- B. Endothermic fusion reactions
- C. Sensible heat glass melt (net)
- D. Wall heat losses
- E. Cooling & leakage heat losses
- F. Flue gas bottom regenerator
- G. Regenerator heat losses

Figure 3 *Typical energy output diagram for industrial float glass furnace producing 600 metric tons molten glass per day. Sensible heat* of glass melt plus reaction energy account for 39 % of the supplied energy. More than 30 % of the supplied energy is exiting the furnace by the flue gas heat contents.*
* Including reheating of re-circulating glass melt. Energy consumption is 6.13 gross MMBTU/short ton molten float glass.

ENERGY SAVINGS BY PROCESS PARAMETER OPTIMISATION

Energy balance models enable the evaluation of influences of possible process or design changes on the energy consumption of glass furnaces. The influence of the following parameters on energy consumption can be analyzed using these models:

- air excess in combustion process;
- batch humidity;
- emission coefficient (radiant properties of combustion space) of flames;
- pull rate or specific pull;
- changes in raw materials, e.g. replacing batch by cullet or limestone by quick lime (CaO);
- batch preheating temperature;
- reducing recirculation of relatively cold melt from refiner or working end back to the melting end;
- regenerator size;
- insulation layers;
- cold air;
- size of furnace & combustion chamber;
- oxygen enrichment;
- type of fuel.

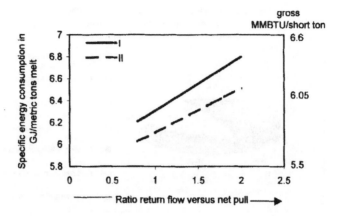

Figure 4. *Influence of return glass melt flow from neck back to melting end of 600 metric*
tons per day float glass melting furnace.
Curve I: forward flow 1370 °, backflow 1165 °C,
Curve II: forward flow 1350 °C and return flow 1180 °C.

Here some results are shown for float glass and container glass furnaces for industry-near
situations. Figure 4 shows the effect of the re-circulation flow of glass melt flowing back from
the working end through the neck into the melting end of a float glass furnace.

The average forward flow temperature (glass melt heading to the working-end) and the
average return glass melt flow temperature, plus the return flow ratio both appear to be
important parameters in the energy consumption of float glass furnaces.

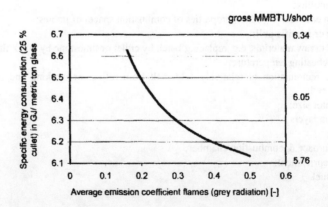

Figure 5. *Energy consumption of float glass furnace (600 metric tons molten glass per*
day, 25 % recycled cullet) dependent on flame emission coefficient
(assuming grey radiation)

Figure 5 shows the effect of the average emission coefficient of the flames covering the glass melt in a float glass furnace on the specific energy consumption. Natural gas flames without soot typically show emission coefficients between 0.15 and 0.18, but fuel oil flames producing more soot may give much a higher radiation emission (0.4-0.5). An increase of the flame emission coefficient is very effective for obtaining energy savings especially at the lower values: an increase from $\varepsilon = 0.16$ to 0.30 gives about 4-5 % energy savings for this float glass furnace.

For a representative modern end-port fired regenerative container glass furnace, energy balance studies have been performed. The most energy efficient end-port fired furnaces show energy consumption levels of 3.3-3.9 GJ per ton molten glass (3.1-3.65 gross MMBTU/short ton molten glass), dependent on the cullet fraction in the batch.
For one of the most energy-efficient furnaces identified by benchmarking, an energy balance has been derived. The energy output distribution is given by figure 6.
Figure 7 shows the calculated effect of air excess and pull on the specific energy consumption of the end-port fired container glass furnace.

Table 2, shows the changes in specific energy consumption for an end-port fired container glass furnace, cross-fired regenerative float glass furnace, oxygen-fired container glass furnace and recuperative tableware furnace by changes in process parameters. *(Negative values of extra energy demand refers to energy savings).*

Well-insulated container glass furnaces show <u>average</u> heat loss of about 4-5 kW/m^2 (furnace walls, crown, bottom) including cooling losses. An increase of the heat losses by 50 % (6-7.5 kW/m^2), due to poorer furnace insulation will result in about 15 % more energy consumption of an end-port fired regenerative container glass furnace.

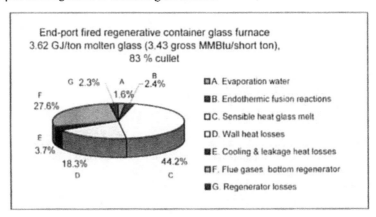

Figure 6. Typical energy out put distribution for end-port fired regenerative container glass furnace. Data: 260 metric tons molten glass per day, 80-85 % cullet, 2-3 % batch humidity, 60 % regenerator efficiency, green glass. About 48 % of the supplied energy is used for the endothermic fusion reactions and to heat the melt.

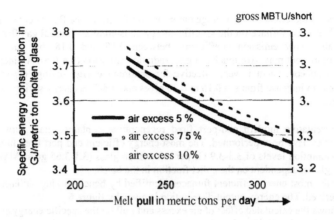

Figure 7. *Effect of pull rate and combustion air excess on energy consumption of end-port*
fired regenerative container glass furnace (80-85 % cullet)
(2.25-3 metric tons per m² per day)

Energy balance modelling shows energy savings of about **4.5-5** % by exchanging limestone
($CaCO_3$) by CaO in a soda-lime-silica batch (**15** mass-% of normal batch is limestone,
complete batch contained 25 % cullet).

CONCLUSIONS ENERGY BALANCE MODELLING

Figures 3 and 6 show that about 27-32 % of the supplied energy is lost by flue gases exiting
the regenerators of end-port fired container **glass** or float glass furnaces. In case of oxygen-
firing, the energy losses by flue gases is also > 25 % of the energy input.

The reclaim of flue gas energy by batch & cullet pre-heaters, as shown by the data of table 2
offers a significant energy saving potential. Today, about 9 furnaces are equipped with batch
preheating systems [7, 8, 10, 11], heated by flue gases from glass furnace regenerators or
oxygen-fired furnaces, showing glass furnace energy savings of 10- 18 %.

Table 2 shows that water in the batch will drastically increase the energy demand of glass
furnaces due to the high evaporation enthalpy of this water and extra energy required to heat
up the water vapour. Each 1 % extra humidity (10 kg on 1000 kg batch) will increase the
energy consumption by 1.5 − 2 %. However, some batch humidity is often required to avoid
batch de-mixing during transport and charging into the furnace and to limit carry-over. Thus,
there is probably an optimum water content, often 3 to 4 mass-% water in the batch is applied.
High combustion air excess or oxygen excess levels are very costly for recuperative furnaces
due to the low air preheat temperatures, and for oxygen-fired furnaces, not only increasing the
energy consumption (about 2 % extra energy for 5 % more oxygen) only, but also the oxygen
consumption. Increasing the oxygen excess from 4.5 to 9.5 % will increase the energy
consumption of an oxygen fired glass furnace by 2 % and oxygen consumption by 7 %.

As long as the glass quality allows an increase in melting pull, this generally will decrease the specific energy consumption of industrial glass furnaces. However, at very high specific pull

Table2. Changa in energy consumption by process modifications

	clear float glass	green container glass	exhaust temperature green container	flint
glass type	clear float glass	green container glass	green container	flint
furnace type	cross fired regenerative	end-port regenerative	oxygen-gas	unit melter recuperative
boosting	no electric boosting	no electric boosting	electric boosting	no electric boosting
fuel	natural gas	natural gas	natural gas	natural gas
pull in metric tons molten glass per day	600	260	210	50
batch humidity in mass-%	3.5	2.5	4	3.5
cullet % (glass basis)	25	83	75	50
regenerator efficiency*	59	60		35
emission flame	0.16	0.16	0.28	0.16
batch temperature in °C	40	40	30	40
combustion air excess %	10	7.5	4.5	10
temperature melt throat or neck (forward)	1370	1340	1320	1300
recirculation versus pull	1.65	0	0	0
primary energy consumption** GJ/ton melt	6.63	3.62	4.46	7.61
gross MMBTU/short ton	6.3	3.42	4.22	7.2
	extra energy demand	extra energy demand	extra energy demand	extra energy demand
emission flame from 0.16-0.20	-2.50%	-1.20%		-2.35%
emission flame from 0.16-0.30	-4.90%	-2.90%		-5.50%
1 % less humid batch	-1.50%	-2.00%	-2%	-1.70%
air or oxygen excess 5 % extra	1.60%	1.70%	2%	4.30%
temperature melt - 10 °C	-2.02%	-1.10%	-0.82%	-1.74%
10 % more pull	-2%	-1.90%	-3%	-2.50%
10 % more cullet	-2.20%	-3.10%	-2.90%	-2.75%
batch preheating 300 °C	-14%	-16%	-20%	-16%
10 % less recirculation neck-melting end	-1.10%			

* ratio sensible heat transferred to combustion air versus sensible heat of flue gas top regenerator (reference T = 0 °C)
** including primary energy required for oxygen and electricity generation

rates (pull per unit area of glass melt surface) the glass quality may be jeopardized or the regenerator capacity is insufficient for effective air preheating and the energy demand may increase at further pull rate increments[5].

Temperature control of the glass melt exiting the furnace and keeping this temperature at the minimum allowable level may decrease energy consumption, but the expected effect is rather small (0.5-2 %). Important is the enhancement of the flame radiation emission properties for instance by promoting soot formation. **An** increase of the (grey) emission coefficient from 0.16 (typical sootless natural gas-air flames) to 0.20, already decreases energy consumption by about –2 to –2.5 % especially for furnaces with high specific energy consumption levels.

Melting of cullet instead of normal batch saves about 29 % energy in oxygen-fired container glass furnaces and 26 % in end-port fired regenerative furnaces according to the energy balance models (for constant pull). This is in close agreement with the observations found in the benchmark studies for container glass furnaces (29 % energy savings).

The use of de-carbonated raw materials such as CaO, MgO or calcium or magnesium oxides gives significant reductions in energy demand of glass melting in E-glass and soda-lime-silica glass furnaces.

The most energy efficient container glass furnaces show energy efficiencies close to 50 % (fraction of energy effectively used for fusion and heating of the melt) and batch preheating can even increase these efficiency levels to values of 50-60 %. The most energy efficient furnaces or fuel are not always the most economic choices: Energy saving measures may require costly investments or energy efficient fuels may be relatively expensive. Thus, apart from an energy efficiency analysis **of** glass melting process, an economic assessment including capital costs, interest rates, costs for air pollution control, fuel and operating costs during the furnace lifetime have to be made to obtain the most optimum economic choice.

Acknowledgement: The author wants to thank Mr. Hans van Limpt from TNO, Mr. Geert Jacobs from GEM-Projects and Phil Ross for their support and discussions.

LITERATURE REFERENCES

1. Beerkens, R.G.C.; Limpt, van H.A.C.; Jacobs, G.: Energy efficiency benchmarking of glass furnaces. Glass Sci. Technol. 77 (2004) no. 2, pp. 47-57
2. Limpt, van J.A.C.; Beerkens, R.G.C.: Energy efficiency benchmarking 2003: Glass Furnaces. TNO Report IMC-RAP-05-11045 (2005).
3. Beerkens, R.G.C.; Limpt van J.A.C.: Energy Savings Measures for Glass Furnaces. Proceedings of the 80th Annual Meeting of the German Glass Society (DGG) 12.-14. June 2006 Dresden, Germany
4. Directive 2003/87/EC of the European Parliament and of the council: establishing a scheme for greenhouse gas emission allowance trading within the Community and amending Council Directive 96/61/EC, 13 October 2003 http://europa.eu.int/eur-lex/pri/en/oj/dat/2003/l_275/l_27520031025en00320046.pdf
5. Trier, W.: Glass Furnaces – Design, Construction and Operation. Society of Glass Technology, Sheffield England (1987) originally from Springer Verlag (German language) 1984
6. Fleischmann, B.: Konventionell beheizte Glasschmelzöfen für die Hohl- und Flachglasherstellung im deutschsprachigen Raum. Teil 2. Glasqualität und Energieverbrauch. Glastech. Ber. Glass Sci. Technol. 70 (1997) no. 3. N27-N34
7. Schroeder, R.W.; Snyder, W.J.; Steigman, F.: Cullet preheating and particulate filtering for oxy-fuel glass furnaces. Proceedings of Novem Energy Efficiency in Glass Industry, Ed. Th. Nohlmans, L. Moonen, and R. Beerkens. Amsterdam, The Netherlands 18.-19. May 2000, pp. 78-84
8. Lubitz, G.; Beutin, E.F.; Leimkühler, J.: Oxy-fuel fired furnace in combination with batch and cullet preheater. Proceedings of Novem conference on Energy Efficiency in Glass Industry, Ed. Th. Nohlmans, L. Moonen, and R. Beerkens. Amsterdam, The Netherlands, 18.-19. May 2000, pp. 69-78
9. Conradt, R.:Pimkhaokham, P.: An Easy-to-apply method to estimate the heat demand for melting technical silicate glasses. Proceedings 2nd International Conference on Advances in the Fusion and Processing of Glass. 22.-25. October 1990 Düsseldorf, Germany, Glastech. Ber. 63 K (1990) pp. 134-143
10. Barklage H.: Erfahrungen mit Primärmaßnahmen zur Verbesserung der Umweltsituation - Vortrag v. d. Fachausschuß VI der Deutsche Glastechnische Gesellschaft DGG (Umweltschutz), 14. Oktober 2004 in Würzburg, Germany
11. Ehrig, R.; Wiegand, J.; Neubauer, E.: Five years of operational experience with the SORG LoNOx® Melter. Glastech. Ber. Glass Sci. Technol., 68 (1995) nr. 2, pp. 73-78

LEONE INDUSTRIES: EXPERIENCE WITH CULLET FILTER/PREHEATER

Larry Barrickman and Peter Leone
Leone Industries

INTRODUCTION

Leone Industries is a family-owned glass container manufacturer located in Southern New Jersey. In 1995 we decided to add another furnace to the factory. In order to meet certain site "bubble" limits on air emissions—and particularly nitrogen oxide emissions— we elected oxy-fuel combustion. In order to meet federal New Source Performance Standards for particulate emissions Edmeston A. B. of Sweden engineered and fabricated for us a system designed both to filter particulates and to preheat cullet. Praxair later acquired the rights to the technology Edmeston employed, which it calls the Praxair Cullet Preheater/Filter System. The purpose of this paper is to share with the industry our experiences both with the Preheater/Filter System in particular and with cullet preheating in general.

TECHNICAL

The Preheater/Filter System derives from an installation partially sponsored by British Glass and installed on a recuperative air-gas furnace at Irish Glass Bottle in Dublin. Its central concept is electrostatic and involves two devices.

IONIZER

Furnace exhaust gases are inducted through an Ionizer [figure 1]—essentially a 4 ft. diameter pipe that contains a high voltage low amperage grid. The grid imparts a corona charge to the particulates in the exhaust gas stream, which at this point has been controlled to a temperature of 580 ° F - 620 ° F by a dilution air fan. The grid is isolated from the housing by way of four insulators that have cooling air blowing on them also from the dilution fan.

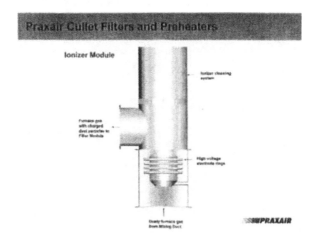

Figure 1: Ionizer Module

FILTER MODULE

The gases thus charged are then inducted through the Filter Module [figure 2], a cross-flow heat exchanger that collects the charged dust particles from the gas and transfers heat from the gas to cullet. The gas is drawn by fan into the center of the Filter Module and radially through a concentric packed bed of cullet about 24 inches thick which flows down through the Filter Module toward the furnace. Immersed in the cullet bed is an electrode cage, suspended from the top of the Filter Module, by which voltage is applied to the cullet to produce a charge of polarity opposite to that with which the exhaust gas has been charged. The particles in the exhaust gases stick to the cullet pieces and flow with the heated cullet into a to a screw feeder where the cullet and associated dust mixes with the batch and is fed into the furnace by way of an oscillating charger. The filter module holds enough cullet-- ≈ 100 ton, to provide the required surface area for collection of the sulfate dust from the furnace and to minimize the pressure drop of the gases flowing through the device. Because of the cross-flow design and the long cullet residence time the exit temperatures of the cullet and of the filtered gases are each about 450° F – 550° F. No higher pre-heat temperature can be achieved in the Filter Module as designed because introduction of gases at a higher temperature than 600 degrees F eventually results in a short-circuit: since glass becomes more conductive as it heats up, a hot spot develops in the cullet beds which draws more and more current until it runs to ground through the body of the Filter Module itself. The filtered gases exiting the Filter Module are blown out the stack. The theory is that the system will: preheat the cullet; filter the gases and return the collected particulate to the furnace.

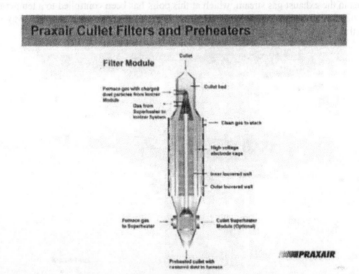

Figure 2: Filter Module

Our requirements—and our hopes—for the Preheater/Filter system differed from that for the Irish Glass installation in several respects. First and foremost we—or should I say the

Environmental Protection Agency-- required that the system reduce the furnace's particulate emissions below 0.2 pounds per short ton of glass produced. Second our oxy-fuel furnace would generate a much hotter exhaust stream than that from the Irish Glass recuperators—and that exhaust would have a higher concentration of particulate and of moisture than that of an air-fuel furnace. Third our installation had to resolve a challenge that revealed itself at Irish Glass: post-consumer recycled cullet contains organic material that volatilizes—sometimes offensively—when cooked. And fourth we set a target pre-heat temperature that was, in retrospect, rather ambitious.

Because of these challenges Edmeston included in the system as designed two other devices—one primarily to deal with the impurities in post-consumer cullet and one to heat cullet bound for the furnace to a higher temperature than the Filter Module's electrostatic systems could tolerate.

PYROLYZER
If a glass plant uses a significant amount of external cullet, removal of organic material is necessary as well as the moisture. The pyrolyzer [Figure 3] is a packed bed, cross flow, heat exchanger that exchanges heat from the hot gasses introduced into a shell around the cullet bed. These hot gases are controlled in a mixing duct to a temperature between 950° F – 1000° F by way of a cooling fan and P.I.D. loop. These gases are pulled through the pyrolyzer by way of a recycle fan (1), and discharged into the downward exhaust flues of the furnace. This fan is also on a P.I.D. loop to control the temperature of the flue at the insertion point to 1600° F to 1650° F. It has been determined that the volatilized organic material in this gas stream will be incinerated in the 1600° F gas stream. The cullet passing through the pyrolyzer is heated to a discharge temperature near 700° F dependent somewhat on the cullet flow through the pyrolyzer and fed into the filter module with some cooling or directly into the furnace. The gas discharge temperature will vary from 180° F – 500° F again dependent on several factors.

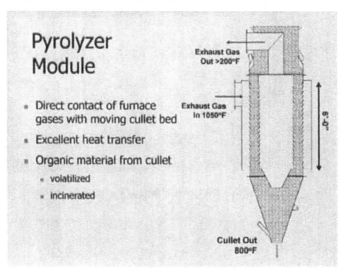

Figure 3 - Pyrolyzer

SUPERHEATER

The final step in the preheat system as designed was to increase the temperature of the cullet in the filter module discharge to ≥ 750o F. Because of the gas inlet temperature restriction to the filter module for optimum filtering and sustained operation, the concept of adding heat after passing out of the grid area was added. The superheater is a short residence time section (similar to the pyrolyzer), which is driven by the superheater fan and discharges the exhaust back into the ionizer mixing duct [See Figure 4].

Fortunately we enjoyed the advantage that the new furnace would be installed in a purpose-built building which could, to a great extent, be tailored to the voracious space—and especially vertical space-- requirements of the preheater/filter and its ancillary devices. The furnace itself was designed so as to melt at least 280 short tons per day without electric boost, although the business plan was initially to install forming machinery that would process from 180 to 220 short tons per day. The furnace and the building to house the furnace, hot end, and inspection equipment were constructed in 1997 and 1998. The furnace was commissioned on April of 1998 and the Filter/Preheater by September of 1998. Until 2002, the pull rate was 180-220 Tons/day. In late 2002, a third forming machine was added allowing us to get closer to its nominal pull rates. Following the material flow on the slide [figure 5] shows how each of the materials eventually gets into the furnace.

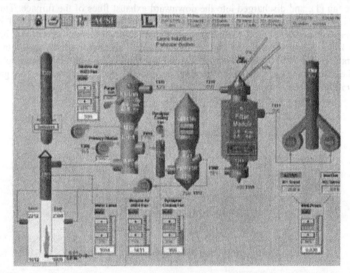

Figure 4

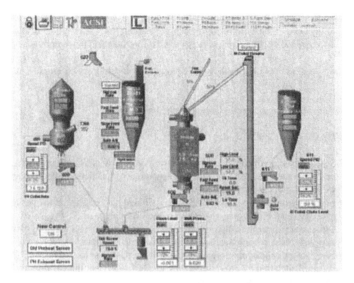

Figure 5

Our experiences generally with the Cullet Filter/Preheater have convinced us that, under certain circumstances, a system of this sort can be a viable means both of filtering the exhaust and of conserving energy in an oxy/fuel furnace. But our experiences have also suggested to us that the design and configuration of such a system—and indeed any cullet preheat system—must be guided as much by several factors that affect reliability as by factors that affect the efficiency of heat transfer. These factors are:

- The abrasiveness, hardness and fragility of cullet. Our experience is that cullet will eventually wear away pretty much any material in pretty much any circumstance. And cullet will, so to speak, wear itself out: the more it is handled the more it will splinter—which creates problems for a system that depends on gas finding its way through a bed of cullet at a predictable rate. So, in general, the less cullet is handled, and the more that handling is on a plug-flow basis, the more likely the system is to be reliable. But even then cullet will eventually wear surfaces away.

- Exhaust from an oxy-fuel furnace is a highly-concentrated dirty gas that has a very high moisture content. Hence it cools off very quickly; as soon as it does substances entrained in the gas will drop out. And some of those substances will drop out if presented with any substantial pressure drop. So the gas-handling portion of the system upstream of the Filter must first be properly sized to the volume and temperature of gases at each point in the system; and having a means of access to clean the gas ductwork is helpful indeed. It should be noted, though, that in our experience the system could usually handle the changes in furnace throughput—and therefore gas volumes—that derived from job changes.

- these two factors complicate, in turn, the choice of materials for use in the system. The combination of high temperature, abrasion and various sorts of chemical attack forces unsatisfying compromises on the materials scientist.

- And, as noted both above and below, system sizing is very important.

Let me now talk about the system's challenges and its successes.

PYROLYZER CHALLENGES

1. Sizing of the pyrolyzer is important. It is a cross flow heat exchanger. The transfer of heat from the flue gas to the cullet is dependent on residence time. In our installation, we feel it should have been larger.
2. Inlet gas temperature also proved to be a critical path because if the gas temperature exceeds $1050°$ F plugging will occur in the lower areas of the pyrolyzer.
3. Outlet gas temperature also is important to prevent moisture and plugging of the gas discharge duct, recycle fan, and the duct work to the flues.
4. We also had to provide a way to introduce air if we lost the recycle fan. We accomplished that by providing a way to add furnace cooling air directly into the recycle lines at the point of insertion to the flues.
5. Reducing the rate of flow of the cullet through the pyrolyzer by splitting the two cullet flows, inhouse cullet directly to the filter module, solved the low gas outlet temperature and increase the cullet discharge temperature by $150°$ F - $200°$ F.

IONIZER CHALLENGES

1. Initially required cleaning 4 to 5 times a week. A water/air cleaning system was installed from the beginning to accomplish this task. The water cycle was abandoned later.
2. The duct work through out the system was designed for a nominal pull rate of 280 tons/day. Before we achieved this in 2002, velocities of the dust laden flue gas was low. Therefore, periodic cleaning of the ducts was necessary.
3. The reasons for abandoning the water cycle were the insulators would crack causing a short or the firing ring pins would get bent or have a hard buildup on them resulting on the ionizer not performing well.
4. The plugging of the duct work to and out of the ionizer required us to put clean out doors through out the system to avoid shutting down the system.
5. The inlet temperature was controlled better after the decision was made to reduce the flue temperature to the stack using a water lance.
Several operational challenges were associated with this device

FILTER MODULE CHALLENGES

1. Blown fuses to the power supply became a problem. Inlet temperatures of the gas to the filter module in excess of $650°$ F would increase the current draw, tramp metal [ferrous/non-ferrous] also caused a problem. We installed along with the existing metal detectors non-ferrous detectors. In addition, on occasion an insulator would break.

2. Installed a limiting reactor to the power supply to reduce the voltage as the current increased. However, this resulted in some times lowering the voltage below our required voltage. To be in compliance we installed an adjustable potentiometer to set the voltage at our limit and now run consistently.
3. Most recently, we had a problem with the electric grid in the filter module. The pipes that make up the grid wore thin in a few areas and broke loose causing a direct short by falling the louvers. This was repaired and plans are being made to change out the entire grid over the Christmas shut down.

SUPERHEATER

We no longer use the Superheater and have settled for the time to stay with the $450°$ F $- 550°$ F discharge temperature from the filter module.

SUPERHEATER CHALLENGES

1. Temperature control and spot heating would cause plugging in the discharge area of the filter module.
2. When using the Superheater the discharge from the Superheater fan would pick up small pieces of glass adding to the weight on the stack testing filter.

OPERATIONAL CHALLENGES

1. We believed at first it was best to continue to wet our batch even though we knew that it would loose some of its moisture when the hot cullet was mixed with it. The batch moisture dropped below 1% by weight and the steam produced would actually spew the dry batch out of the charger hopper and onto the furnace crown.
2. The dry batch also would flow freely into the furnace when the oscillating charger would move to a new position. We now run the charger in a fixed position.
3. Shaker/Separator proved to be more trouble than it did good. It became a high maintenance device. Separating the two cullet streams also fixed this problem.
4. Any one change in the operation of the filter/preheater would affect several other components in the system and would require some time to equilibrate.

CONCLUSION

I have just briefly explained to you the operation of the Edmeston/Praxair Filtering System. In addition, I tried to give you a sense of the problems associated with a new technology development. There is a good side to our experience with this system. The most notable benefits are listed below.

Benefits
1. The results of the two stack tests had different pull rates can be seen in Figure 6
2. Figure 7 shows pull rate in tons/day on the bottom and fuel usage expressed in mBTUs/ton. Figure 8 shows energy consumed with pull rate corrected to 260 tons/day and at 50% cullet.
3. Allows for future increase pull if needed.

4. Batch costs are lower on this furnace than our other furnace because of the collection of the sulfates and selenium the latter costs has increased by 200-400%.

STACK TESTING RESULTS

Pollutants	September 1999 Pull rate 296 T/day	September 2001 Pull rate 267 T/day	NJ DEP Regulations
Particulate Matter	0.04 g/kg	0.06 g/kg	
	0.08 #/ton	0.11 #/ton	0.20 #/ton
Sulfur Dioxide	0.33 g/kg	0.44 g/kg	
	0.65 #/ton	0.88 #/ton	
Nitrogen Dioxide	0.65 g/kg	0.53 g/kg	
	1.30 #/ton	1.05 #/ton	5.5 #/ton
Carbon Monoxide	0.005 g/kg	0.01 g/kg	
	0.01 #/ton	0.02 #/ton	
VOC	0.01 g/kg	0.01 g/kg	
	0.02 #/ton	0.02 #/ton	

Figure 6

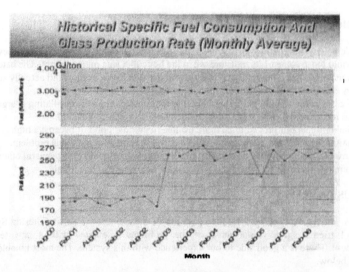

Figure 7

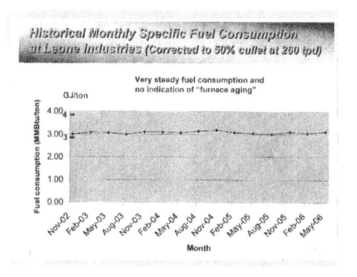

Figure 8

PETROLEUM COKE TECHNOLOGY FOR GLASS MELTING FURNACES

M. A. Olín, R. Cabrera, I. Solís, and R. Valadez;
Vitro S.A., San Pedro Garza García, NL, Mexico

INTRODUCTION

In order to maintain competitiveness, the glass industry faces the challenge of developing technologies and operational practices that guarantee "low energy consumption" and / or "low energy cost" production processes, while reducing operational costs and complying with all the environmental regulations within the territories were it operates.

For many years the Vitro preferred fuels for glass melting operations have been natural gas and heavy fuel oil. However, in Mexico, the price for these non-renewable fuels has become highly volatile and has been increasing since year 2000, making the search for alternate fuels a major priority for the company, and within this context, the seeking of the use of petroleum coke as a possible solution to the problem.

Petroleum coke is a solid waste by product of the petroleum refining industry and its production is projected to continue to increase, due to the fact that many refining plants are undergoing or planning process conversions in order to obtain higher value products; light fuels.

Between May 2000 and December 2001, Vitro set up test facilities and undertook a series of pilot tests in order to demonstrate the technical feasibility of using petroleum coke as an alternate fuel for the glass industry. Several laboratory and industrial tests were carried out during this period with successful results.

A new combustion technology process was developed for three types of regenerative glass furnaces: glass containers end port and side port, and float flat glass side port. Pilot tests confirmed the technical and economical feasibility of the use of petroleum coke and paved the road for the industrial implementation.

During the development of these tests, several technologies and specific applications were developed, such as: 1) supply & logistics; 2) fuel grinding and handling; 3) fuel reception, transportation and storage; 4) solid fuel firing; and 5) environmental pollution control systems.

Since the beginning of the project, continuous financial valuation studies were developed to determine the economical viability of the endeavor. For this purpose, a very detailed valuation model was configured and used in parallel to the technical efforts; investment and cost analysis data has been gathered, upgraded and updated at the different stages of the project life. Between 2003 and 2005 projects to implement at industrial level the whole set of technologies in the first operation plants where developed.

BACKGROUND

As everybody is aware, in the last decade the worldwide glass industry has been very active searching for new technologies to reduce the energy consumption for the glass melting process. A good amount of resources have been dedicated by many companies to develop projects aiming to find process improvements to current melting technologies and / or a totally new melting process that can result in a signicant energy consumption reductions compared to the current ones.

In the period between the last months of the year 1999 and the first of the year 2000, the Vitro Mexico glass melting plants were suffering a significant operations cost increase due to the natural gas unexpected price increase and volatility.

In May of 2000 the former Vitro Operations President asked to the Corporate Technology Office to work together with the Corporate Energy Office in a preliminary theoretical study to determine the technical feasibility and economical viability of the use of petroleum coke as an alternative fuel to natural gas in the glass melting furnaces.

During June of 2000 the Team developed the study and concluded that there were good probabilities the pet coke usage in the glass melters could be developed successfully and the potential economical benefits look very attractive. Based on this, a Development Project Proposal was prepared and submitted for approval. The Proposal included the development for both type of furnaces; end port (containers) and side port (float glass), and proposed what we called a prototyping "FAST TRACK" R&D approach instead of a Lab R&D one. In August of 2000 the Proposal was approved and the Project was launched.

DEVELOPMENT PROJECT

The project activities were divided in three large areas: containers end port pilot furnace, float glass pilot furnace, and research and development subjects. The first two areas were done in parallel and included developing all the engineering required, equipments manufacturing, building & installation, and operations launch. The third area was also done in parallel with the other two, and was divided in five main subjects: combustion, refractories, glass quality, environmental, business model work teams for each of the three areas were integrated and during 16 months project activities were developed with successful results.

In May of 2002 project proposals to develop three pilot plants: containers, float glass, and grinding, were prepared and submitted for approval to Vitro Executive Board. The three projects were approved, and from August of 2002 to December 2004 were gradually developed and launched. Currently, the three pilot plants are totally equipped and partially operating with pet coke.

During the last two years ongoing process improvements & equipment optimization activities have been done. In this period of time the learning process has been intense and sometimes overwhelming, facing innumerable technical challenges and organizations resistance to change obstacles. Managerial Factor such as having a sponsor from top management, who has delivered the idea and the vision of the use of pet coke as a fuel, his involvement reviewing progress activities and also establishing drivers to the organizations to pursue the objective to have low cost alternative fuels for future operations were key for the project to move on and not stop when facing barriers, but finding ways to overcome every situation.

Other non-technical challenges faced by the team members were to find experts' advice with know how and knowledge of specific areas to address an application with no previous information available, since few attempts had been made in past and with no success. As an example, equipment such as burners and pet coke feeder had to be designed considering the actual conditions that glass operation require and no relevant information was available for the use of a powder fuel in glass.

PET COKE COMBUSTION

Combustion was one of the main difficulties faced during the development stages. Firing a solid fuel is not as easy as operating with natural gas, since it has to be stored in specific containers for the fuel and also transported in a regular and controlled manner towards the furnace. These two operations required at the beginning the installation and testing of new equipments and also to establish new setting parameters to operate them and process development looking for optimal performance.

Pneumatic conveying principles resulted key during the first stages of development, it was not only important to burn completely the pet coke, but it was also important and even more difficult to keep pet coke flowing at constant rate to transport it towards the burners that made it necessarily to get deep understanding in another areas of knowledge not so common in glass industrial practices.

In order to feed a constant rate of fuel to the furnaces, we tried at least three different very expensive dosing commercial systems (designed for other industrial applications) that didn't worked as required, and to make the story short, we ended up developing a completely new pet coke feeder that complied with all the process requirements and that could make the investment project cost-effective.

Commercial feeders had way up capacities compared to the glass furnace requirements, so scalability was the first factor we had to overcome. Another important factor was the high price of the commercial systems that will erode the benefits of using a low costs fuel if we would choose to operate with them. But most important, was the fact that none of the tried systems operate in an appropriate way neither in fuel consumption rates nor in the operational conditions of pipe lines, burners and supplied air of pneumatic conveying.

The new Vitro feeder was simpler, less expensive, and overall operated within the fuel rates required by the different capacities of the glass furnaces and operational conditions of the feeding designed system.

Additional challenges were faced also regarding pet coke burners. For one type of furnace we tried some commercial burners with reasonable results. But, as progressing in the development stages we had to make several changes until we come out with a burner that operated with optimal conditions and that was also found to be the best suitable for that specific furnace.

In another different type of furnace, we started with an outsourced idea of a pet coke burner, that did not work in the very beginning, but after several stages of changes and adjustments, we ended up operating a new redesigned burner with a more appropriate flame configuration and moreover a reasonable well fuel combustion.

Combustion of pet coke is not an easy task. Pet coke requires lots of air, a good mixture of combustion air, time and also good burner geometry to burn completely. In this regard, configuration of port and the interaction with the burner was a key factor to reach a complete combustion.

Fundamental tools such as simulation modeling were constantly used in all the development stages. Having good models that approach combustion and fluid dynamics of solid fuels in pneumatic conveying was also very important. It was also key to "test" different burner configuration and the interaction between fuel and combustion air, considering the port design and internal conditions of a glass furnace, that the only way to cover a great amount of possibilities in a short period of time was by simulating them. Mathematical modeling has been **a** factor that reduced developing time since a good approach can be reached before having to try anything on the field.

REFRACTORIES

Another important issue since the very beginning was the refractory behavior when firing pet coke in the glass furnace. The pet coke is a solid waste. It contains some materials as impurities to deal with. **As** it may be understand firing pet coke will have some influence on refractories, and as it normally happens, there is no free lunch. In the case of firing pet coke it was found that with proper material selection and some maintenance furnace practices, pet coke will operate with no major hassles to refractories as the regular fuels currently used in glass industry.

Effects on refractory will vary of course due to the furnace type, operation temperatures, refractory materials quality, maintenance practices, and additional operational influences, such as, carry-over and volatilization that are common in the glass industry; and these components vary significantly from factors such as furnace design, temperature and operational conditions. In other words, the same type of factors **as** when operating with a regular fuel.

There are still some issues that are being addressed, hut as mentioned, we believe that with proper material selection and additional maintenance practices, the use of pet coke might represent a future option for other glass makers, as well.

GLASS QUALITY

Glass quality has been a key component to make operation with pet coke feasible. Prior to any production tests, a glass quality assessment was performed addressing the main parameters that it was believed the pet coke would have influence on them. Aspects such as glass color, seeds and faults, density and other properties were considered. This first assessment showed that minimal effects would be expected in the product quality if certain operational conditions were maintained when operating with pet coke, and with that in mind, trials were started at furnace level with pet coke combustion to track all the glass quality parameters.

Since the very first trials glass quality was tracked in all the different aspects for the final products and no change was observed in them. Of course, some adjustments in operational settings had to be done in order to maintain the same levels presented in the glass production with other different fuels.

As today, several furnaces have been operating with pet coke and a countless number of technical challenges have been faced, and good amount of adjustments have been made. Process development is built through incremental steps versus quantum leaps, and in this case it was not the exception. So, trying a different fuel required sometimes thinking out of the box, and also getting some knowledge from other areas than glass industry and that were not quite familiar to us that made it even more challenging.

In terms of energy, after operating for several years some furnaces with this fuel, we have not seen major changes in terms of thermal efficiency of the furnaces (below figures). However, it is important to say that flames of pet coke are more radiant that natural gas flames since they come from the combustion of a solid fuel and soot creation generates more radiative energy than when firing a gaseous fuel.

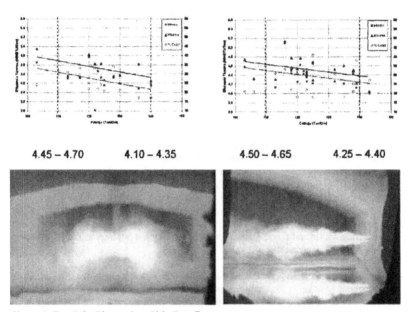

| 4.45 – 4.70 | 4.10 – 4.35 | 4.50 – 4.65 | 4.25 – 4.40 |

Figure 1. Pet Coke Flames in a Side-Port Furnace

ENVIRONMENTAL SYSTEM

When the information about the usage of petroleum coke was understood for all, the necessity to find a system to control flue gases emission was clear due the high sulfur content. At the same time we start a search for the best way to riddle with the sulfur content in three ways, prevent to enter, control in the process or at pipe end. Taken this new requirement we developed a search around the world for companies that could help to design this system.

This search lead us to some good amount of companies, large, mid and small ones, and we found out that the interest of large companies was null because they like to work on large projects like power generation facilities or refineries; and mid and small ones were a good number and with different technologies and emissions control results offered.

On the other hand, the standard technology used for many glass companies throughout the entire world are base on dry systems and / or electrostatic precipitators (EP), and no much around the sulfur content, mainly due the extensive use of natural gas as clean fuel. EP is a solution with a large capital investment and also requires a large print foot near the furnaces.

Then the search was re-directed to find a suitable technology that could fit with our requirements: low capital investment, excellent sulfur removal, minimal operational problems, and same time could solve the particulates emission problem.

In the beginning, the system we used was an experimental one with a large vessel and plastic tanks, to hold liquids for the scrubbing of the gases. After several trials we found the operation window where the system worked in tandem with the furnace and combined well for the internal furnace pressure control.

Because the system was totally new for us, at beginning we had a lot of bad and hard experiences like plugging of drain lines and scrubbing nozzles, agitator shaft mechanical failures, plastic piping and metal tanks run off, others. However, with a lot of work through several years and with much patience from our top management all problems were solved and with a good amount of lessons learned documented and included in our engineering manuals.

The system requires equipment to separate the solids from liquids (vacum filter in our case) in order to recover the water and minimize the water consumption, especially in Monterrey that has water restrictions by the climate and region characteristics (semi desert). Also we use recovered water from drainage to minimize the use of fresh water, and this allows us to use more water than if it came from a fresh supply.

The vacuum filter works well but initially we had problems with the very fine particles and also with the set up of the rotation and fabric installation, it was tricky but not impossible to solve and works well at the end of the second day.

Having a new process to understand and the need to make it work was a challenge to all of us, so we began approaching new ways to operate it and short cuts to handle new situations; like a sudden shut down of the gases extractor fan and the furnace continuing burning fuel, causing a furnace pressure very high increase. However, the time spent in the tedious construction and analysis of the Design Failure Mode and Effect Analysis (FMEA) was a blessing and payout, because the prevention for emergency stops of the system worked beautifully and avoided major problems.

Another interesting experience was the usage of the concrete stack to exhaust the flue gases with the new system in one of our furnaces. After the gas scrubbing process the gases are wet and saturated, but the condensation occurs early than anticipated in the tall stack and then we had vapor going out of ventilation holes and the stack look like the stack was going on fire, putting all the plant people on alarm. This and other additional problems convinced us to change the system and spend the money in a new parallel stack, now our furnaces had twin stacks and depends of the fuel used which one is in use.

Similarly, with tall stack we learned that it is possible to switch from one fuel to another having a minimum temperature differential; the draft can be enough to start the pull of the flue gases in seconds with out problems and now our large units can switch between fuels in automatic mode with out problems. The wet flue gas treatment system is typically located after the regenerator's chambers with a deviation from the tunnel stack to the new system. The cleaned gas is discharged to the new stack.

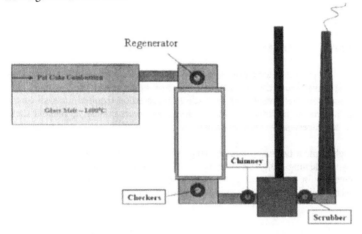

Figure 2. Typical arrrangment

Selection of reagent is important to take care to improve the efficiency of removal and maintain operational cost down. The system operates reaching 95 % removal of SO2. confirming our idea to use such equipment and technology in the project.

With the pass of the time we have been able to develop a better system and improve the performance of the unit to have more reliability of the operation and longer times between maintenance, for example; keep the inside of the scrubber vessel cleaner avoiding solids deposits and incrustations. The bottom line for the scrubber system was to keep a low cost for the flue gases cleaning; investment and operation. After all this time, after a lot of experience gained with the new system, and with a good amount of resources spent, now we can say the technology works and can be easily transfer.

Advantages:
- Well – established technology
- removals of 95% are common
- SO2, NOx and Particulate emissions comply with the current Mexican Regulations
- Reagents used by the process are low cost, plentiful and readily available
- Waste is not a hazardous one

The system can be operated by glass furnaces easily and is not sensitive to operating modes such cycling or reversal of firing.

Finally, after this new and exciting experience we continue working in how to improve the environmental system, we are convinced that it can still be optimized; saving more space, investment and operating cost for our furnaces. The best condition for the improvement is a fresh look to unusual situations and problems, looking to break old paradigms that limit our vision and search for new ways to do things.

CONTROL SYSTEMS & AUTOMATION

The overall scope for this project in the automation area was to assure the appropriate interaction between the control systems involved in the different stages of the process in order to carry out a synchronous operation altogether, transferring the process information required for each control system and making the process information available to higher levels in order to make it useful for top management and supervisory control systems.

Changing fuel in a glass plant became a real challenge specially because it was not a common fuel for the industry, so dealing with a powder fuel required the knowledge of different topics like pneumatic conveying in its different ways, new types of sensors, as well as new control strategies; all of them, with different degrees of integration into the existing Melter Control System.

The main guidelines followed to begin the development of these control systems involved in the new process are listed below:

- Low cost
- Integration & Connectivity
- Flexibility
- Reliability
- Concurrent engineering development
- Project design in stages
- Safety issues

Then, focus on the basic control equipment that would full fill the previous requirements and following a good strategy to split the process into natural control areas according the process flow. The results of the previous steps derive in the following scheme (Figure 3).

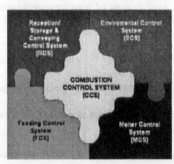

Figure 3

The main challenge was to define a vertical or horizontal integration along the plant because the nature of each control system requires both directions, so at the end we did a mix getting excellent results in the overall performance.

In order to have a simple integration to the existing furnace control system, we worked on a flexible solution that would allow us to interact between several applications at different communication levels. First, we work on the top down design of the overall architecture, and then focus on all the interactions required by the systems involved in the different process stages, and next on specific solutions for each system. The big issue was to work on all this process design concurrently.

One of the most critical decisions on this project was related to the communication network. When we started this project six years ago some of the industrial networks available today, that we know for sure are very reliable, were not released completely at that time. The available ones, were very complex and expensive to implement due to the type of integration that we wanted to achieve, so we worked very close with suppliers and their technical staffs in order to be sure that we were on the right track; and we did it right, up to know we haven't experienced any big issue regarding communication networks.

As I already mentioned, interactions between different control systems at different levels, starting at the lowest level, direct hardwired links and ending to the highest level; high speed communication links played a very important roll for the entire application. Special attention and consideration were taken on safety issues and set as high priority for the control system design because of the type of process being handled, as well as the integrity of local control and the data being transferred and used by other controllers.

According to the previous scheme, four new control systems were designed and integrated into the glass melting process: Reception/Storage & Conveying, Feeding, Combustion and Environmental. The melter control systems are the ones already in use by the plants so we just made minor changes in order to integrate the new control systems.

The Reception/Storage and Conveying Control System consists of receiving the pulverized fuel from the outside and transferring it in a safe way towards the storage silo where it remains according to the fuel consumption from each furnace. Several sensors as well as its own controller are used in order to meet the safety issues and perform automatic control.

The Feeder Control System performs the feeding of the pulverized fuel to the furnace taking care of several control variables in order to keep a very stable fuel feeding during the firing cycle. This is a multifunction machine and was developed specially for this purpose.

The Melter Control System is the one already in operation; this represented another challenge because of the type of control system already installed. At this point and knowing that there were different control systems in the plants, we decided to develop a very flexible architecture that allow us to integrate the different technologies used in the other control systems in order to make a transparent integration, and trying to avoid getting into the Melter Control System as less as possible. At this point the Combustion Control System came into the scene as the main tool use for this purpose, easy integration for all systems, it doesn't matter the kind of controller already installed, with the proposed flexible architecture it was easy to carry out the integration using a direct hardware link for safety issues and a hi level communication link.

Figure 4 shows a general block diagram representing the Combustion Control System. As we can see there is a very close interaction between all the control systems involved.

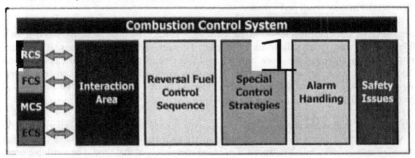

Figure 4. A general block diagram representing the Combustion Control System

In case of a failure of this control system, we developed several control strategies in order to keep the overall process operating. This system has been in operation for several years and one of the must amazing events was when we fired three different fuels in the same furnace. We fired gas. oil and pet coke in a float side port glass furnace.

One of the most important things that we found out regarding the combustion of the pulverized fuel into the glass furnace is the need of a very high stability over the pulverized fuel, as well as the conveying air. That means the right sensors for the controlled variable as well as the actuators for the manipulated variables, and a set of different advance control strategies in order to keep a very tight control. As we know, the big issue on regenerative type furnaces is the reversal, so every time a reversal firing cycle takes place, a disturbance in the process is generated. taking this in mind, we developed several control strategies in order to minimize this disturbance in the firing process.

Testing with different equipment like valves. actuators, sensors, etc, was a key issue in order to have reliable equipment that could stand the operating conditions, perform according to the process requirements and generate the field information required for the control system in order to manipulate the process accordingly.

One of the major constrains faced during the development of this project was that not all the equipment required was available in the market, or used by other industries for this kind of fuel. in order to solve the process issues a lab was built and equipped with different instruments in order to determine the key variables needed to monitor and control the process, as well as, to define the required specs for the equipment. This stage was very important because in the design process. cost, was a very important driver for the project in order to make it feasible.

At this stage we complete the elements to perform the fuel firing using a pulverized fuel in a glass furnace, then: there was the need to develop a complete new process and its associated control system in order to clean the exhaust gases before sending them to the atmosphere. The main challenge for this control system was designing and developing a new control strategy able to manipulate the exhaust gases generated inside the glass furnace without disturbing the internal pressure.

The first approach was to integrate the available technology in the market. We did the implementation with an external company but we didn't have the expected results in our process, so we decided to get involved directly in this process and use most of the existing equipment; making several changes in some parts of the process and rebuilt the entire control system concept. After several months, we achieved a very big improvement in the operation, but still having problems with the reliability of the field equipment, as well as, with some of the key operations from the unit. We focused on industrial applications as a reference in order to use similar principle of operation for the measurement and control of the key variables, and we tested different kind of sensors until we found the equipment that allowed us to have a reliable operation. Once we had a more stable operation we went into an optimization stage of the equipment adding strategic sensors that allowed us to have additional safety issues and process information about the performance of the critical elements. Now we can tell that with all these elements we have achieved a more stable operation as well as less down time and a very stable internal pressure control of the melter.

COAL GASIFICATION

John Brown
Glass Manufacturing Industry Council

TNTRODUCTION

The sustainability of the global economy and population relies on an abundant, stable and affordable energy supply. Today, that economy puts its faith in the continued accessibility to OIL.

Since Oil's introduction 150 years ago into our industrialized economy, society has reaped the benefits of this high energy and low cost energy Source Oil extraction has steadily climbed to meet the growing needs of the energy hungry world.

Until now. 2006 has been designated by some' as the PEAK year. The PEAK year is the point in time when the steady increase in oil extraction plateaus and after a few years, a little less oil is extracted each year. The **US** remains the dominant consumer; requiring 25% of global output. Further complicating oil's future is the rising demand of India and China whose rapid industrial growth will lead to tremendous supply-demand swings; affecting oil pricing worldwide.

The world needs an alternative energy source. Developing one will not be easy. Development will require capital. Fortunately, the **US** is blessed with an abundant supply of coal that can be developed through chemistry to supply our gaseous and liquid fuel stocks. To avoid the pending energy supply gap to the Glass Industry we must be at the forefront of this energy revolution to ensure our furnaces have a reliable and cost effective supply of energy.

For the past three years, the Glass Manufacturing Industry Council (GMIC) has been investigating this possibility. In late 2003, GMIC representatives visited the TECO ICGG utility plant in Tampa, Florida. Following this visit we met with industry reps and experts in the gasifying field. The meetings culminated in the drafting of a Glass Industry "White Paper" detailing the opportunities inherent in the clean coal gasification processes. The White Paper suggests a process to determine optimum alternatives for our industry. **A** gasification "task force" was created to guide our pursuit of this and other alternative fuel resources.

GMIC has also been involved in meetings organized by both industry and government sectors.
- The DOE, with support from the National Energy Technology Laboratory, is investigating gasification technologies most suitable to industry.
- The consumers Alliance for Affordable Natural Gas (CAANG) is supporting efforts to develop gasification **as** a means to reduce demand for the diminishing supply of Natural Gas.
- Other initiatives involving public/private collaboration are investigating a range of alternative fuels such as biomass, pet-coke, black liquor and municipal solid waste.

---

[1] **Richard Heinberg.** *The Party's Over:* New Society Publishers, 103-104 (2003)

A wide range of feedstock products could be developed to provide sustainable combustible materials to the glass industry. Identifying and validating the supply is in its early stages. We will likely see new ideas in the future as natural gas prices fluctuate. The likelihood of rising gas prices will lead to new technological developments. The GMIC will stay on top of these, and in turn, keep the glass industry informed.

> *My father rode a camel. I drive a car. My son flies a jet airplane.*
> *His son will ride a camel.* -- Saudi saying

DEMAND FOR ENERGY

The global economy is heavily influenced by the available and cost of energy. The price of natural gas is a threat to the US manufacturing base. The two primary reasons the price of natural gas did not rise above $15 per million Btu this year was due to a combination of mild winter and the decrease in manufacturing as three million relatively high paying manufacturing jobs were permanently shut down or idled over the past five years. Industrial "demand destruction" is responsible for freeing up 0.8 trillion cubic feet of natural gas or 9% of US industries total natural gas demand. A study of the seasonal cyclic costs of natural gas over the past four years reveals a trend where prices drop as storage systems are filled in preparation for winter and summer gasoline blends shift to winter formulations. For five months, prices decline but the corresponding peak is inevitably higher than the previous year's peak.

Quite simply, the US faces a serious natural gas supply-demand crisis. Since 1999 the rapid increase in price of natural gas has cost consumers an additional $200 billion: not including the costs related to reduced employment or increased prices of electricity.

In 2004, the "supply gap," defined as the amount of natural gas the US receives from Canada and imported LNG, has increased 42% to 3.7 TCF. The predictions for the US natural gas supply are in question. Canadian exports have decreased and LNG has shown only modest increases. Our manufacturing base is reluctant to invest in the US due to the erratic price swings of natural gas and other forms of energy when compared to more stable pricing available elsewhere.

The stabilizer for the cyclic bottom in natural gas pricing is the huge demand for natural gas to drive turbine generators. Over 235 thousand Megawatts of new power has been added in the last decade. Today nearly $100 billion of the $140 billion installed power generating base sits idle because the price of natural gas exceeds the price of the electricity produced. This back log in generating power provides the rock bottom price to which natural gas can fall before excess supply is consumed to produce megawatts. About $7 per million Btu, because once reached turbine generators are brought on line and begin reducing the supply side of natural gas. Of all the uses for natural gas, Electric Power generation shows the highest potential for growth.

Improved energy efficiency across all sectors of the economy should be a high priority. Peter Garforth compares US energy efficiency to the best practices in Europe. Manufacturing is not so bad. We employ engineers and invest in good practices so our energy per unit of output is only 20% higher than EU best practices. Our automotive industry is moving ahead but if pushed with a serious consistent government message could have done much better than the 40% disadvantage per passenger/distance metric used for comparison. The real opportunity is in our homes and here is an opportunity for glass insulation, and flat glass for improved thermal efficient windows. On a per unit of area for similar climate, our homes and buildings require

150% more energy. In addition, 70% of the power generating industry is consumed in our homes and buildings. This is a real glass business opportunity and an area where our industrywill add REAL security to the US population.

This could be the next big political statement. Thomas Friedman in The New York Times, October 13, 2006[2] spoke of a call from James Carville who coined the phrase "It's the Economy, stupid". "Energy Independence", he said, " It's now the No. 1 national security issue." If one party would take on energy and say we will do this to make the US independent of foreign energy, and make a believable case, they would win the popular vote. The public doesn't want gasoline or Btu tax. No, they want government to impose much higher auto mileage standards and more stringent energy codes on BUILDINGS AND APPLIANCES. People want a tough regulatory response. Reducing dependence on foreign energy is voters' top national security priority. Our nation deserves a better energy policy than, "Praying for a cool summers and a warm winters."

All options must be on the table. Consumers win when energy markets compete between and among the energy supply options. The US has a marvelous gift from nature in their coal reserves. While coal gasification has been practiced for perhaps 150 years, many significant improvements have only recently been developed. IGCC (Integrated Gasification Combined Cycle) is a superb technology that also provides a solution to environmental challenges not met by traditional powder coal plants.

Some facts about our supply of coal:

- The US has nearly 19 billion tons of recoverable coal reserves
- 27% of the world's coal reserves are in the US

When combined with shale, pet coke and our remaining oil reserves, we have a total equivalent of 1.9 Trillion Barrels of oil. Nearly triple the total of all the Middle East oil reserves of 685 billion barrels. Last year the US extracted 1.1 billion short tons of coal representing 22.6 quads of energy. This equals the annual energy requirement of US industry. Only three companies represent 40% of all the coal mined.

Later in this discussion the quad will be mentioned again. Because the quad is such a large number, it needs further explanation. Peter Angelini compared a quad of energy to a stack of oil barrels that reach from earth to the orbit of the moon. At that point you have stacked a quad of energy. The US consumes nearly 100 of these stacks of energy each year, and all US Industry, 22 of these columns of energy. The glass industry requires only 25% of one quad.

THE COAL GASIFICATION PROCESS

Our glass industry is a big consumer of natural gas (225 trillion Btu or a quarter of a quad) but only represents 1% of the industrial energy consumption with chemicals taking over 30%.

---

[2] Thomas L. Friedman, The Energy Mandate, *The New York Times,*Oct. 13, 2006 OP-ED Columnist

If energy independence is important to our industry then we must embrace the opportunities inherent in coal. Coal gasification produces clean fuel. When combined with oxy-firing of glass furnaces, oxygen-blown gasification provides the required available heat and low emissions.

GMIC has visited two DOE sponsored project sites. In Tampa Florida the Tampa Electric Company supplies 250 Megawatts from 2200 tons of coal, per day. The second, the Wabash River project in Indiana produces 260 MW from 2400 tons of coal per day or 2000 tons of pet coke. The construction cost of each plant was $438 million with 50% of the funds from DOE. An additional coal, oxygen blown plant was visited following The American Ceramic Society's Glass and Optical Materials division meeting in Greenville, SC this spring. Eastman Chemicals was gracious and provided a tour and answered all questions. Their operation is for providing feed stocks for making chemicals. Without the gasification plant, Eastman Chemicals could not compete with cost of natural gas in the US.

What is Gasification? Gasification is the conversion of solid and liquid hydrocarbons, in the presence of oxygen and steam by partial oxidation into a reducing environment gas stream. This stream is comprised primarily of about 50/50 hydrogen and carbon monoxide.

Advantages include:

- Greater efficiency, 20% increase (33% to 40%) in realized energy from a unit of coal for the newer Integrated Gasification Combined Cycle (IGCC) process
- Economic capture: mercury emissions with the potential of CO_2 capture plus the basis for competitive production of high quality liquid fuels
- Byproduct markets
  o Coal Gasification: H_2 and CO with some CO_2 that can be removed or further processed into fuels
  o Electricity production: HS and ash are trapped by the cooling process when steam turbines produce power. These can be separated and sold. The resultant ash or slag is a fine shiny black material with chemistry similar to glass. This material is non-soluble in water and about 200 mesh

The final coal gasification process is termed Methanation. It has the disadvantage of a 25% energy penalty but if you must substitute for natural gas with no loss in thermal energy, this is an alternative.

Peter Walsh and I took the available data from the Wabash plant and performed an energy balance.

- 2,090 tons of pet coke produces 2.49 billion Btu per hour (2.5 million CFH equivalent natural gas) at 85% efficiency from the coals heating value
- Output is 96MW from steam turbine at 33% efficiency and 192 Mw from the gas turbine at 40% efficiency
- 36 MW are required to operate the plant. Of this, 20 MW are required to separate oxygen from air which requires about a ton of oxygen per ton of bio-mass
- Net energy to the grid is 252 MW

The glass industry would run this plant a little differently.

- The glass industry would run the high efficiency syn-gas at 85% efficiency and burn in an oxy-fired glass furnace with an efficiency of 54% (combined 46% efficiency)
- Rather than provide 252 MW to the grid we would take 192 MW from the gas turbine and save the $75 million for that turbine and have 1.64 billion Btu/hr; enough energy to operate eight major glass plants

IGCC plants are clean and offer fossil fuel without environmental baggage. Heating values of air blown compared to oxygen blown systems and natural gas are 150, 290 and 1028 Btu/CF, respectively. If, as Donald Bonk says, "I can ever get John to forget all the propaganda on slagging-gasifiers and that syn-gas is only H_2 and CO, the possibilities are endless." For example, Donald Bonk reports a biomass indirect gasifier using a propriety catalyst recently produced energy results of 499 Btu/SCF.

We have considered three models for the Glass Industry:
- The first was a modular designed unit to supply the needs of a 300 tpd container furnace. Three modules would handle a 600 tpd day float furnace. NETL offered to design the module during one of our earlier discussions contributing ideas toward our alternative fuels white paper
- The second alternative is a larger plant built to serve several energy intensive industrial sites within close proximity

- The third option is to invest up to $2 billion in a plant built on top of a coal mine utilizing syn-gas for methanation. Even with the 25% energy penalty, methanation has the advantage of direct integration into the national pipe lines distributing natural gas. Investors could tap natural gas and be independent of oil and natural gas pricing

A contract was to be signed by October 1st 2007 and I had hoped to report on this novel approach to replace some of the shortage in natural gas. While I can't say anymore it is an attractive approach to replacing natural gas with an identical energy substitute.

Outside the US, gasification is booming. Sixty six (66) GW were being produced world wide in 2005. Chemicals utilizing 40%, only 20% in the US. Source of biomass is 45% from petroleum residuals like pet-coke and 45% from coal and about 10% from natural gas.

In Chicago, GTI operates a state-of-the-art gasification pilot plant. Vann Bush directs the gasification R&D facility and made a presentation at the September 19th GMIC board meeting we held at GTI. Vann Bush has accepted GMIC's proposal to include a demonstration of syn-gas as the fuel with oxygen as part of our experimental design for the submerged combustion melter. The pilot plant furnace will be a demonstration that syn-gas can be used to melt glass.

An important consideration is the ratio of syn-gas to oxygen. With syn-gas the ratio is 3.4 units of fuel per two units of oxygen whereas natural gas is utilized with a ratio of one unit of fuel per two units of oxygen. Same volume of oxygen per unit of time to produce the same heat input.

The two DoE IGCC projects are big in energy consumption of 2.5 billion Btu/hr and cost of 438 million dollars.

In Genoa Italy, HMI Inc. supports a laboratory gasifier producing between 8-9 million Btu/hr in Genoa, Italy.

Perhaps a glass company will want to operate their own gasifier--but it is unlikely. The smaller gasifiers are less reliable and less efficient. Uptime of 85% is seen as a drawback. Eastman Chemicals solved the down time issue by building two gasifers; relying on the 2nd as an expensive backup.

Contained within the 2005 US Government Energy bill is funding for industrial projects. I have summarized several pages into a single slide.

- Title 1 is for clean coal power initiative and has a cost of Federal Incentives from fiscal year 2006—2015 of 2.5 to 3 billion dollars

- Title 2 is of interest to the glass industry and details support for industrial gasification. Eligibility require the ability to gasify coal, bio mass or petroleum residues for applications related to seven industries including *Glass!* Total cost of Federal Incentives FY2006—FY2015 $1.4 billion

SEQUESTERING CARBON DIOXIDE

William Shakespeare, 1564-1616, inhaled air containing 280 ppm of CO_2. Today, with each breath we draw 380 ppm of CO_2. No one today can claim to fully understand the consequences of the increase in CO_2 but human kind is running an uncontrolled experiment.

Insurance is an ultra conservative institution. The largest, Swiss Reinsurance, is offering credits for CO_2 sequestering. World insurance revenue was $3 trillion in 2005[3], about a third more than total sales of the oil and gas industry. At the same time US insurers paid out $71 billion this past year, or 14% of net premium earnings. For the past 20 years, payments had averaged 3.3% of premiums. Climate change is increasingly recognized as an on going significant global environmental problem, says Irv Menzinger, head of Swiss Reinsurance Co's sustainability and emerging risk management department.

The world's largest re-insurers, by premiums are offering to underwrite carbon emission credits for businesses to change the global greenhouse gas equation. Insurance is in the business of making money. To remain profitable they are willing to participate in the reduction of greenhouse gases. If areas like the Gulf Coast are more vulnerable, insurance premiums will increase or insurance will not be offered. Insurance companies will remain profitable.

Hurricanes produced a result that made the Gulf Coast appear it had been bombed. The heart of U.S. petrochemical production as well as oil and gas was affected. Consumption was reduced almost as much as production in the early weeks. Still it led to $15 per 1000 cf natural gas in December, despite extremely mild November. It will be two to three years and large expenditures required to rebuild infrastructure. This reduces funds for new development.

---

[3] Christopher Hinton, *Insurers see Global Warming as Threat,* Wall Street Journal, Wednesday, September 27, 2006, Market Place Section.

Would you want to invest in Gulf of Mexico deepwater projects?

The atmosphere has been the dumping ground for exhaust from tail pipes and smokestacks because it is easy and cheap.

The good news is that the technology for capture and storage already exists and that the obstacles hindering implementation seem to be surmountable.

Oxygen blown coal gasification's inherent advantage that the bulk of the exhaust gases are CO_2 and water vapor; which are easily separated. By eliminating nitrogen from the oxidation stream, the volume of exhaust gases is greatly reduced. In addition, the syn-gas produced has little nitrogen and will combust in glass furnaces with minimal NOx, as occurs in today's oxy-gas furnaces.

While the global economy is focused on large power plants which represent 25% of the world's carbon dioxide emissions, all large energy intensive industries need to plan for sequestration.

A new large 1,000 MW coal fired generating plant produces six million tons of CO_2 annually. This is equivalent to the emissions of two million cars.

Researchers[4] believe that the best destinations in most cases for CO_2 capture will be underground formations of sedimentary rock loaded with pores now filled with brine. To be suitable, the sites typically would lie below any source of drinking water, at least 800 meters. At this depth ambient pressure is 80 atmospheres, high enough that the pressurized injected CO_2 is in a "super-critical" phase—one that is nearly as dense as the brine it replaces in geologic formations.

At the Krechba field in the Algerian desert, BP engineers are mapping the geologic layers via seismic echolocation soundings to determine where best to place the CO_2 well. Potential places for sequestering carbon dioxide show up in color in these 20 meter thick reservoirs. In this form the CO_2 can be expressed in barrels. Each year a 1000 MW plant will need to inject about 50 million barrels of super critical carbon dioxide—about 100,000 barrels a day. After 60 years of operation about three billion barrels, or half a cubic kilometer would be sequestered beneath the surface. Some 500 giant oil fields exist in this size range and have produced about two thirds of the 1,000 billion barrels of oil the world has produced to date. The CO_2 can turn a profit in these old oil fields to boost production of oil.

In the Algerian Salah gas project plant natural gas contains too much CO_2 for commercial use so the excess is removed by chemical absorbers, compressed and injected into a brine formation two kilometers below the surface.

The next few years will be critical for the development of carbon dioxide capture-and-store methods. Policies to make these practices profitable may come from the insurance industries and a desire by a determined government to join with the world's leaders to work to reduce global warming. Meeting clean air-water and land requirements can be profitable; GE has said that

---

[4] Robert H. Socolow, *Can We Bury Global Warming?*, Scientific American, July, 2005, 49-55

Green is Green. We made great strides in meeting objectives of reducing NOx, SOx and particulates in the 80's and 90's. There is hope that CO_2 pollution can be mitigated as well.

And now to mention one last regional opportunity. John MacKay has an operation south of Columbus, Ohio where he produces 76,000 pounds of carbon nano-fibers annually. His process requires sulfur and carbon to grow the fibers. Hydrogen is the byproduct. What an interesting set of needs. His operation would produce an equal stream of hydrogen and methane; eight million cubic feet per year. While this represents a fraction of the needs of a large furnace, with planned growth by several orders of magnitude, this could be an ideal match for industry to provide a new raw material as well as a clean high energy fuel. Michael Greenman and John Chumley visited this plant to see if it fits with Alfred Universities' approach of capturing hydrogen in glass micro-spheres for automotive fuel.

RECOMMENDATIONS

I requested a list of the top developments that would enable coal gasification to become competitive. Gary Stiegel, Technology Manager-gasification NETL, provided the first six items though Donald Bonk. *(Note: Italicized print represents billion dollar plus opportunities)*

1. Small economical and efficient gasifiers—2.5-100 Mw scale range
2. Robust dry feed systems for handling mixed feed material, e.g., MSW/coal, biomass/coal; improved feed preparation systems *(opportunity for raw material suppliers to the glass industry)*
3. Multi-contaminant control technologies (process intensification) for sulfur, NOx, Hg, Chlorides, ammonia, etc. *(opportunity of emissions control suppliers to the glass industry)*
4. Low cost air separation unit for medium Btu gas (operating at ambient conditions) for chemical applications and if CO_2 capture is required (otherwise air blown is fine) *(opportunity for our air separation industries that supply oxygen and nitrogen to the glass industry)*
5. Approaches to economically eliminate refractory lining in gasifiers *(opportunities for the refractory supply industry)*
6. System modularity and standardization
7. Reduced oxygen needs. Today most systems use one ton of oxygen per ton of coal. Some processes can cut the required oxygen in half. *(Oxygen minimization is a huge opportunity in cutting costs and hopefully our air separation supply industry will jump on this one.)*

CONCLUSIONS

- Growing participation by the glass industry in developing the "white paper" but now must show concerted, collaborative effort
- No individual company has the potential to develop it on its own
- We must work collaboratively as we are doing on the submerged combustion melter project

Additional Recommendations
- Become an active member of GMIC's Gasification Task Force
- Assign an individual with the support of their company
- Commit resources to achieve our objectives, with the support of state and federal government agencies

ACKNOWLEDGEMENTS

Donald Bonk

Vann Bush

Peter Garforth

Michael Greenman

Sho Kobayashi

Rolf Maurer

Donna Ransom

William Rosenberg

Gary Stiegel

Peter Walsh

Andy Weissman

PREHEATING DEVICES FOR FUTURE GLASS MAKING, A 2ND GENERATION APPROACH

Ann-Katrin Glüsing
Zippe Industrieanlagen GmbH, Wertheim, Germany
Email: k.gluesing@zippe.de

ABSTRACT
Cullet and batch preheating systems can theoretically be installed at any existing glass melting furnace with greater than 50% cullet in the batch. Preheating of only the batch has been problematic and is not considered as proven technology. Up until now, the integrated add-on technology of waste heat recovery through batch preheating and oxy-fuel combustion has not managed to really break through unto the market in spite of the joint benefits in emission control and operating costs. Batch preheating has not found acceptance due to long pay-back periods based on energy savings alone and the lack of long-term commercial experience. This lack of experience has prohibited the industry from evaluating potential process benefits, such as increased pull, better quality, and increased furnace life, which can greatly increase the technological financial attractiveness for conventional air-full furnaces. The idea of enhancements to this preheating technology towards a second generation of batch and cullet preheaters is presented.

1. INTRODUCTION

Globalization, as a result of exogenous factors, has to be recognized by energy intensive and labour-intensive sectors as the glass manufacturing industry in Europe. The glass industry lost half its workforce between 1980 and 2003. Productivity in 2003 was 80 % higher than in 1991, and this upwards trend is expected to continue[1]. Unrestricted capital flow, modern communication and modern logistics, e.g. low-priced imports are threats for glass production in an enlarged Europe.

Future glass furnaces must produce a higher glass quality with higher thermal efficiency and lower emissions of CO_2, NO_x, SO_x, and particulates in order to remain competitive in Europe. They must be able to recycle exhausting stream condensables and particulates, and be able to incorporate large percentages of glass cullet into the final glass product. Significant strides in future glass making processes have been made by the introduction of oxy-fuel combustion as an alternative to the air-fuel combustion.

The next step in this development is the introduction of cullet preheating, which promises significant cost reductions from energy savings and process benefits. Because the cullet is preheated, the required energy flux to the melt will be reduced. Thus, the firing rate can be reduced, reducing NO_x and CO_2 emissions. Increasing pressure to improve glass industry competitiveness relative to alternative materials, e.g. plastic packaging, while at the same time decreasing furnace emissions, makes introduction of the granular bulk solids preheating logical an important next step in glass evolution.

---

[1] The challenges facing the business related service sectors are described in The "Communication on the Competitiveness of Business related services and their contribution to the performance of European Enterprises, COM (2003)747

2. STRUCTURAL CHANGES AND ENVIRONMENTAL NEEDS IN AN ENLARGED EU 25

Due to the paradigm shift of a globalised production society towards an innovation society the achievements of innovation head starts by intellectual capital and material capital is a necessity for securing competitiveness of production "Made in Europe". The renewed Lisbon strategy[2,3] set as a goal for the European Union to become the most competitive and dynamic area of the world by 2010. In an enlarged EU 25, the convergence of countries has to be seen in relation to the gross domestic product GDP[4] and the national innovation capacity[5] and, in general, with the need for harmonisation of environmental standards on the road to sustainable production[6]. This plan promotes new and best performing technologies among technology push and market pull to bridge the gap between successful demonstration of innovative technologies and their effective introduction to the market to achieve mass deployment.

The EU has a set of common rules on permits for industrial installations. These rules are set out in the so-called IPPC Directive[7]. IPPC stands for integrated pollution and prevention and control. In essence, the IPPC Directive is about minimising pollution from various source points throughout the European Union. All glass manufacturing installations covered by the Directive are required to obtain an authorisation (permit) from the authorities in the EU countries. The permits must be based on the concept of Best Available Techniques (BAT). In many cases BAT means quite radical environmental improvements and, sometimes, it will be very costly for companies to adapt their plants to BAT.

Faced with regulatory and competitive pressure to control emissions and reduce operation costs, as in 2005, CO_2 emissions have become a market-based instrument through emission trading schemes, taxes and charges, glass manufactures are considering a variety of options for reducing the overall energy consumption. The Commission, therefore, urges Member States to make greater efforts to meet the October 2007 deadline by which all installations covered by the IPPC Directive must possess their permits. Further work would be useful in assessing techniques to be considered that for energy usage asked for the EU commission for the upcoming review of the BAT 2007. As a consequence, the improvement of new information about long-term use of oxy-fuel firing is needed in order to assess issues such as the effect of oxy-fuel firing on refractory life and the direct costs of technique with regard to the balance between energy savings and the cost of oxygen for the changes in the glass BREF document.

Technology challenges related to thermal efficiency of glass furnaces are pointed out by the Commission[8], because the waste heat is not fully utilized. Heat recovery systems, as well as preheating systems, e.g., SCR, SNCR, reburning and cullet/batch preheating require improvement.

[2] Implementing the Community Lisbon Programme: A Policy framework to strengthening the policy framework for EU manufacturing: - toward a more integrated approach for industrial policy, COM (2005) 474 is based on a systematic identification of sectoral opportunities and challenges.
[3] New structures for the support of high quality research in Europe, A report from a high level working group constituted by the European Science Foundation ESF to review the option of creating a European Research Council, ESF, 2003
[4] Sachverständigenrat zur Begutachtung der gesamtwirtschaftlichen Entwicklung (2005) and Federal Government of Germany (04/2005)
[5] MIT Technology Review 05/2004, Deutsche Bank Research and M. Porter, S. Stern. Global Competitiveness report 2003/04
[6] On the road to sustainable production, Progress in implementing Council Directive 96/61/EC concerning integrated pollution and prevention control, COM (2003) 354 final
[7] Report from the commission to the council and the European Parliament on implementation of the Directive 96/61/EC concerning integrated pollution and prevention and control, COM (2005) 540 final and EIPPCB, Reference document on best available techniques in the glass manufacturing industry, 07/2000, 12/2001
[8] CPIV – IPPC WG Revision of the Glass BREF, 2005

3. OBJECTIVE

To go beyond regulatory requirements is chance to change the current practice of feeding these raw materials called batch, along with crushed waste glass and rejected ware called cullet, into the furnace at ambient temperature. The overall objective of this study is to demonstrate the technical viability, system reliability, and economic benefits of indirect bulk solids preheating installations as add-on technology for improving the overall economics of furnace glass production in an oxy-fuel melting system in contrast to a conventional air fuel system. Herewith, the main objective must be to preheat the predominantly largest amount of the batch because of the shortage of cullet [13] as a main raw material of container glass in Europe. The proposed program will accelerate batch preheating introduction into the glass industry by performing one full-scale commercial demonstration in the EU 25 compared to a semi-technique lab application. This integrated approach of batch and cullet preheating with oxy-fuel firing is unique because of the combination of both in there infancy state of existing technologies for a non-container glass application.

By quantification of the process benefits, it is anticipated that the application to conventional air fuel furnaces will follow. To avoid infant mortalities as in the eighties, special emphasis is given to the prevention of sticking of material or stoppage of the batch flow as well as the increase of particulate emission of dry batch at furnace entrance next to the aspects of handling of cold and hot bulk solid mass flows. The potential of the higher temperature levels of waste gases at oxy-fuel combustion are especially studied under aspects of design, influence on the melting behavior, accompanied by a heat and energy balance, and even more concerning technological limits and feasibility.

4. PROCESS HEATING

Waste gas heat losses are an unavoidable part of operating any fuel fired furnace. Preheating is a well known technique in the petrochemical sector since the late eighties of the last century. In contrast to the oil industry, that glass industry does not fullfil high fuel efficiency. Modern glass furnaces efficiency only reaches about 50 % as shown by Beerkens[9], primarily due to the heat losses through the stack. Process preheating plays a key role in producing glass and manufacturing products made from these materials. Preheating the incoming raw material is an excellent way to improve the overall glass melting economics[10].

Glass manufacturing is an energy intensive process. From a principle point of view, the glass melting furnace is a chemical reactor[11]. It is designed to convert a batch of raw material, e.g. sand and a wide variety of natural and synthetic materials, such as sodium carbonate, feldspar, and calcium carbonate into a refined, homogeneous and workable melt. This shall be accomplished at a high efficiency of energy exploitation in the shortest possible time and at the lowest possible ratio of output and reactors size. The majority of glass is produced in natural gas-fired furnaces, where raw materials are heated by radiation from combustion gases of existing burners firing over molten glass pool and from furnace enclosure.

As the batch is heated, chemical reactions take place in the decomposition of batch materials and in the formation of a homogeneous melt under release of CO_2. While reducing the demand of natural resources, the use of cullet accelerates the melting process, thereby improving the efficiency of heat exploitation. In container glass melting in Europe, recycled cullet has become the main batch component, as shown in Figure 1. Presently, a cullet ratio of 30 to

[9] Beerkens, R. 1st Glass Manufacturing optimization forum and increased enterprise profitability, TNO, IPCOS 2004
[10] Zippe, Pabst, Rothenmüller, Wiesbaden, 1992
[11] Conradt, R., The glass melting process – treated as a cyclic process of an imperfect heat exchanger. Advances in Fusion and Processing of Glass, 35-44, Rochester, 2003, ISBN 1-57498-156-0

90 % is reached depending on product, type, glass color, availability, and on the level of impurities imported into the material via the public collection system[12].

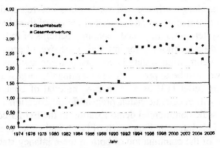

Figure 1. Glass production versus cullet recycling in the German container glass production (1974-2005) in mio. t[13]

The optimization of the melting process per tons of glass in the year 1970 to 2003 shows the increase of the glass recycling ratio from 10 to 87 % and the decrease of the CO_2 emissions from 700 kg to 300 kg per tons of glass [11]. The CO_2 output is reduced in direct proportion to the energy input (≈ 0.2 kg CO_2 per kWh), and beyond this, to the reduction of primary materials (≈ 0.2 kg CO_2 per t of primary raw materials). In the same time, the energy demand of 2600 kWh decreases to 1100 kWh per tons of glass achieved by the German glass industry.

In the energy benchmarking study performed by Beerkens and v. Limpt[14] this value meets the present 10 % level of the furnace ranking (10 % have a lower energy consumption and 90 % have a higher energy consumption) in Figure 2.

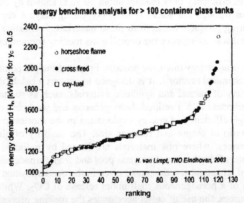

Figure 2. Furnace ranking of over 100 container glass tanks and 10 % level (10 % have a lower energy consumption and 90 % have a higher energy consumption)[15]

[12] Glüsing, A.-K., Observation and dissolution kinetics, supported by microscopy. Advances in Fusion and Processing of Glass, Rochester, 2003, p. 107-115, ISBN 1-57498-156-0
[13] GGA, Gesellschaft für Glasrecycling und Abfallvermeidung mbH, Ravensburg, 2006
[14] Beerkens, R., van Limpt, H. Analysis of energy consumption and energy savings measures for glass furnaces. 80. Glastechnische Tagung, Dresden, 2006
[15] Beerkens, R., Van Limpt, H. Energy Efficiency Benchmarking of Glass Furnaces. 62nd Conference on Glass Problems, Illinois, 2001

In this energy benchmark study, the equivalent specific primary energy demand (GJ per tons of glass) is based on the power plant efficiency of 40 % (1 kWh requires 9 MJ fossil fuel combustion energy), 0.4 kWh electricity for 1 m_N^3 pure oxygen generation and normalised on a cullet ratio of 50 % in the batch. All furnaces should be more energy-efficient than the furnace ranking on the 10 % level (10 % of the furnaces have a lower energy consumption and 90 % a higher energy consumption). During the past 80 years, the energy demand was lowered by a factor of 5, the production power of the furnace increased by a factor of > 15 (as reflected by the pull rate), and the furnace lifetime increased from less than 1 year to more than 10 years [12].

This in itself is an impressive story, which is due to continued improvement in many areas comprising the quality of refractory materials, and the introduction and optimization of heat generation from the hot off-gas.

5. WASTE HEAT RECOVERY
Generally, the aim of the optimization of a glass tank to save energy and meet the environmental should be the on the reduction of the exhaust losses by e.g insulation, even on furnaces that operate properly. Most common technologies for extracting heat from the exhaust gas and recycling it to the process to achieve high temperatures for the combustion air to reach high flame temperatures are regenerators, recuperators, and boilers. Currently, alternatives are under development to reduce the power consumption for oxygen separation by membranes, preheating of the combustion gases, or regenerative as well as recuperative heat recovery systems for oxy-fuel-firing systems as shown by Kobayashi[16], Beerkens[17], and Bos[18]. The thermo chemical regeneration is a worthwhile alternative to conventional physical heat recovery.

Regenerators and comparable systems recover flue gas waste heat to achieve high temperatures of the combustion air. With this high flame temperatures can be reached as well as an enhancement of the heat transfer in the combustion space together with a reduced energy demand. Providing high flame temperatures necessary for the glass melting processes can also be reached by the oxygen-fuel combustion systems, where highly enriched oxygen (98 % O_2) is used in place of preheated air.

Due to the oxidant's low nitrogen content, NO_x generation is significantly reduced compared to conventional air-fuel combustion. Also, the reduced combustion gas volume associated with oxy-fuel combustion lowers gas velocities inside the furnace and stack, which reduces the rate of entrainment of fine batch particles. Heat transfer to the glass is also improved with oxy-fuel combustion. This is, in part, due to the higher flame temperatures achieved and to the heat loss reduction obtained through using fewer and smaller furnace penetrations for the oxy-fuel burners.

A major glass industry focus has been the introduction of oxygen-enriched combustion as a means of reducing furnace emission. Oxygen-fuel combustion reduces a glass furnace's pollutant amount, especially NO_x, while providing the potential for reduced fuel utilization. The gains achieved in energy utilization through oxy-fuel combustion, although significant, in most cases do not offset the oxygen's added cost. Although a lot of work is being done to lower the oxygen cost, present day oxygen costs are equal to or slightly higher than the fuel

[16] Kobayashi, S., Wu, K.T., Switzer, L., Fuel Reduction by Combining Oxy-Fuel Firing with Batch/Cullet Preheating. Glass Problems Conference Energy Workshop, 2005, Champaign, Il.
[17] Beerkens, R., Muysenberg, H.P.H. Vergleichende Untersuchungen von Verfahren zur Energieeinsparung bei Glasschmelzöfen. Glastechn. Ber. 65 (192) Nr. 8, s. 216-224
[18] Bos, H.T.P. De Thermo-Chemische Recuperator. KleiGlasKeramiek (NL) 7 (1986), nr. 6 pp.123-126

costs for most plants. From 351 furnaces the estimated ratio of oxygen fired furnaces is given with 12 % of the total installations in Europe [9].

Here, the synergy approach of cullet and/or batch preheating and oxy-fuel firing focuses on the high temperature levels between outgoing flue gases of oxy-fuel-melters (1350 °C) to preheat the bulk solids up to 400 °C. Batch and cullet exchange with the flue gas permits recycling of otherwise inaccessible flue gas enthalpy of 20 to 40 % with an energy saving potential of 10 to 20 %.

The development of a second generation of preheaters „PRECIOUS" has to be seen as a highly difficult construction and optimisation **task** in order to find the right conditions and parameters in between two unpropitious working conditions: very hot, small volume flow – relatively cold, large volume flow; otherwise **no** increase of efficiency will be achieved.

The small mass rates are at the mean time a challenge for the working conditions of a preheater installation. The direct input of the hot oxy off-gas in the preheater would be an advantage from the point of view of the thermo dynamic availability (exergy) requiring material adoptions and reducing material load, when preheating the cullet and/or batch up to 400-600 °C, instead of 350 °C with the first generation of preheater installations. Next to this, it is important that the temperature of the granular bulk solids does not exceed Tg around 500 °C. The product to be preheated, neither cullet nor batch reach the maximum entrance temperatures of the flue gases, with this the cullet remains lumpy.

Although the elimination of dilution air should be the objective for the second generation of preheaters, the addition of cold air can be necessary because of material load of the preheater. Low mass flows cause low flow velocities; hence, low heat transfers coefficients. This must be solved with regard to the layout of the profiles and dimensions of the waste gas channels in the preheater.

The preheating of cullet and batch at oxy-fuel combustions represents a challenge, due to the oxidant's low nitrogen content. Recuperative waste heat recovery would be technicly feasible, but inefficient with regard to the small volume flow. For that purpose, the following estimation is carried out:

The combustion of 1 m^3 natural gas (here simplified presented as CH_4) needs 2 m^3 oxygen opposed to 3 m^3 flue gases. The heat capacity of oxygen is furthermore about a factor of 0.7 lower than the flue gas existing practically only of polar molecules.

Compared with air-fuel-conditions, 10 m^3 air are opposed to a flue gas amount of 11 m^3 (m=10/11); the heat capacity of the air is around a factor 0.9 lower than the off-gas itself. Hence, the following physical limits of an air or oxygen preheating are given:

- **Air preheating**

$$\frac{m_{Air} \cdot c_{p,Air}}{m_{offgas} \cdot c_{p,offgas}} \approx 82\ \%;$$

- **Oxygen preheating**

$$\frac{m_{oxy} \cdot c_{p,oxy}}{m_{Oxy-offgas} \cdot c_{Oxy-offgas}} < 50\ \%.$$

The low heat capacity of the oyxgen volume flow demonstrates the inefficiency of the oxygen preheating because of the physical reasons (physical limit < 50 %).

Additional problems with recuperative concepts can occur when steel is in contact with hot oxygen. At regenerators, due to the small volume flow creating high investment costs caused by the refractories in a regenerator presenting 60 to 70 % of the refractories in total at a glass tank, inefficient.

The volume flow of the oxygen is even at the same temperature and an endlessly large regenerator not capable of incorporating more heat out of the waste gas. Faced with the problem of optimisation; the quantification of energy flows depends on the layout of the preheater. The conventional regained heat of the flue gas $H_{regenerative}$ in Figure 3 is conveyed to the combustion space of the tank.

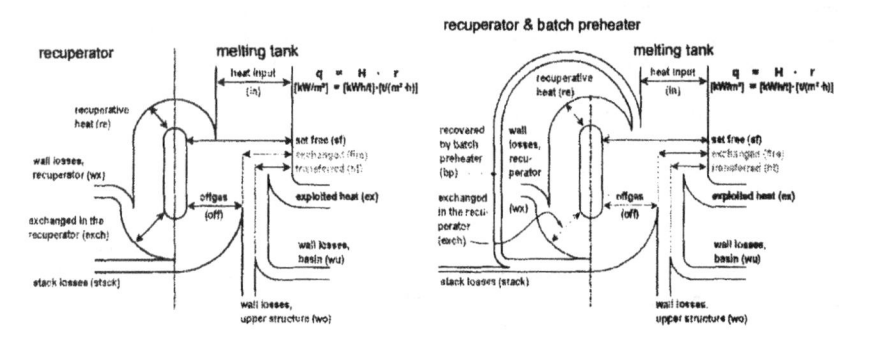

Figure 3. Comparison of a conventional waste heat recovery and integrated with an additional preheater at glass furnaces.

The maximum possible part of $H_{regenerative}$ being transmitted in the melt is

$$H_{max} = H_{off} \cdot (1 - \Delta T_{Stack} / \Delta T_{off}),$$ ΔT_{Stack}, transfer to the stack at end of process, ΔT_{off}. at exit of the tank, after dilution or cooling on 600 °C, and before entrance in the preheater,

which is significant minor upon $H_{regenerative}$. Obviously smaller to the $H_{regenerativ}$ of the conventional recycled waste heat is the recovered amount of waste heat with the preheater $H_{batch\ preheater}$. Because of the lack of transmission loss, during retake in the glass bath the net effect is much higher than the effect of the single contributions.

Assuming a fixed end temperature of 400 °C for the cullet by preheating, than the heat content of the cullet is 380 MJ or 106 kWh per tons of cullet (Sharp and Ginther). A good running tank has a specific primary energy demand of 1150 kWh per tons of glass (10 % level of the energy benchmarking by Beerkens). With 100 % cullet use, an energy saving of 10 % can be achieved, because the energy is not transmitted by the fuel, but direct over the cullet in the melt. The value of 4.1 GJ or 1150 kWh per tons of glass is revised in terms of cullet content of 50 %, the electro energy, and the oxygen generation.

The objective is a cullet ratio of 70 % at the full-scale application, this is 74 kWh, leading at least to an energy saving of 6.5 %, when the cullet is preheated to 400 °C without further additional cost for extra devices, as e.g. steam generating devices.

From the flue gas 190 MJ per tons of glass have to be transmitted, meaning that 190 MJ or 74 kWh can be recovered in the process, when it's succeeds in preheating the cullet up to 400 °C. Looking at the transfer of the heat content per tons of molten glass:

H_{off} = 187 kWh as a temperature level of 600 °C; whereof 74 kWh are to be transmitted on the cullet at a cullet ratio of 70 %. As for the heat transfer

$H_{transfer} = H_{off} \cdot (T_{out}-T_0) / (T_{in}-T_0) = H_{off} \cdot (275 °C) / (575 °C) = 89$ kWh, already showing a successful match at 300 °C due to the preheating of the cullet.

6. BATCH AND CULLET PREHEATING TECHNOLOGY

Indirect and direct batch and cullet preheaters have been developed and installed by GEA/Interprojekt (direct heating), Sorg (direct heating) and Zippe (indirect preheating). Also, a combined direct cullet preheater and electrostatic precipitator was developed and installed by Edmeston. Rotary kilns with a large footprint have only moderate heat transfer rates. Fluidized beds cause an increase of particulates and volatiles in the waste gases and for, an upgrade at existing melters, the required space for the add-on technology is difficult. Both technologies as yet optimized and this has therefore prevented their glass industry application. The available systems are described below.

Edmeston EGB Filter
The edmeston electrified granulate bed filter system is a hybrid system between an electrostatic precipitator for dust removal and a direct cullet preheater. The hot waste gas enters the top of the system and passes through an ionising stage, which imparts an electrical charge to the dust particles. The gas then passes into a bed of granular cullet, which is polarised by a high voltage electrode. The charged dust particles are attracted to the cullet where they are deposited. The cullet is in a shaft and is constantly added at the top and removed from the bottom. The preheated cullet (up to 400 °C) and the attached particulates are charged into the furnace. Applications are limited by the high investigation and maintenance costs but considering a high thermal efficiency.

Direct preheating
This type of preheating involves direct contact between flue gas and raw material (cullet and batch) in a packed or moving bed in a cross-counter flow. The waste gases are supplied to the preheater from the waste gas duct behind the regenerator. They pass through the cavities in the preheater, thereby coming into direct contact with the raw material. The outlet temperature of the cullet is up to 400 °C. The system incorporates a bypass that allows furnace operation to continue when the preheater use is either inappropriate or impossible. Direct preheaters are developed and installed by Interprojekt (formerly by GEA). At Sorg an alternative segmented melter concept, the LoNO$_x$-Melter is combined with a direct preheater. Batch and cullet falling freely through the exchanger is the concept of the raining bed technology at Tecogen. Although the concept shows a good thermal efficiency, the solution has to be assessed under the additional particulate load of the flue gas by direct contact of material and waste gas.

Indirect preheating
The indirect preheating is in principle a cross counter flow, plate heat exchanger, in which the material is heated indirectly. It is designed in a modular form and consists of individual heat exchanger blocks situated above each other. These blocks are again divided into horizontal waste gas and vertical material funnels. Next to the reduced maintenance capacity of the modular build-up, the system incorporates a bypass that allows furnace operation to continue when preheater use is either inappropriate or impossible. In the material funnel the materials flows from the top to the bottom by gravity. Depending on the throughput, the material

reaches a speed of 1-3 m/h and will normally be heated up from ambient temperature to approximately 300°C. The flue gases will be let into the bottom of the preheater and flow into the upper part by means of a special detour. The waste gases flow horizontally through the individual modules. Typically, the flue gases will be cooled down to approximately 300 °C. The indirect cullet preheater is developed by Zippe. Here, an optimization of thermal efficiency is seen, but there are no contaminations of the flue gases through the separate cross-counter flow of material and flue gases while indirect preheating.

7. "INFANT MORTALITIES" OF THE FIRST PREHEATER GENERATION

In contrast to the alternatives gained by boilers, other melter concepts or TCR, this study aims at determing the benefits of the further development of the add-on technology of indirect preheating of cullet and batch. Operational problems concerning the reaction of hot with cold batch components are seen in flowability, charging, dedusting, and water migration. This enormous lack of experience is the reason that almost none of the three integrated batch/cullet preheaters at oxy-fuel-melters in the world of, in total, about 10 integrated installations[19], are still operating. Some problems are discussed in detail below.

For container glass application, the development of the energy costs still show no rentability for the oxy-fuel firing in Europe. Enormous dust development and the charging of the hot melting material have been serious problems at the Gerresheimer melter[20]. With this, the direct Interproject batch/cullet preheater has been modified with a filter module. The same dust problematic, made the installation of an extra filter module necessary at the Heye installation [31]. The energy savings are determined on 25 % and the reduced energy demand is given by 3140 MJ per tons of glass. The described energy savings are fuel savings excluding the generation cost of the oxygen for the combustion. Significant differences for the energy demand during operation with or without the preheater device have been registered. For the same installation Kobayashi[21] specifies the energy demand with 2.9 GJ/t. This strengthens once more the secure identification and comparability of values upon energy saving as performed in [15]. As one of over 100 OI installations, the formerly Gerresheim plant was closed in 2005.

The high dust load as in [20] is also quoted at the Nienburger preheater concept. The Interprojekt preheater was developed at Nienburger Glas formerly, today Rexam in Germany. In consequence of the high dust load a filter module is added, which increases the payback period. Less corrosion next to lower susceptibility through to stoppages in spite of a higher dust load and additional filter modules lead to the installation of the direct preheater solution at Rexam. Jeschke[22] reports of a pull rate of 3.0 t/m²·d without boosting. With electrical boosting, the pull rate can be increased up to 3.5 t/m²·d. Further progress up to 4.0 t/m²·d, by the use of a direct preheater, is possible. Barklage[23] achieved an energy demand of about 3470 MJ/t at cross-fired regenerative tanks combined with a batch and cullet preheater for a cullet ratio of 71 %. Normalized at 50 % cullet as in [14] the energy demand has an average of 3750-3800 MJ per tons of glass.

[19] Report from the commission to the council and the European Parliament on implementation of the directive 96/61/EC concerning integrated pollution and prevention and control, COM (2005) 540 final and EIPPCB, Reference document on best available techniques in the glass manufacturing industry 07/2000, 12/2001

[20] BMBF-Förderkennzeichen 01RV9631/8, Glasschmelzwanne mit externer und interner Vorwärmung des Aufgabegutes zur weiteren Schmelzenergie- und NOx-Reduzierung, 1996-1998

[21] Kobayashi, H.; Wu, K.T.; Switzer, I.; Martinez, S.; Giudici, R. CO2-reduction from glass melting furnaces by oxy-fuel firing combined with batch/cullet preheating. Proceedings XX ATIV Conference "Modern Technologies and techniques for Glass Manufacturing". 2004, Parma, Italy, 171-174

[22] Jeschke, R. Rexam (ehem. Nienburg und PLM), pers. Mitteilung, 2006

[23] Barklage-Hilgefort, H.J. 3 Jahre Betriebserfahrung mit einer querbeheizten Regenerativwanne mit Gemengevorwärmung. Vortrag vor dem Fachausschuss II der DGG, 10/1998 - (today Rexam)

Today's quality and customer demands require the highest cleanliness guidance along the process chain from the hot to the cold end, when a container glass producer is, at the same time. a bottler. as for e.g. Quinn glass[24]. Thus, the pressure on the glass producers with regard to grocery regulations increases the importance of the avoidance of the dust.

Today, the 3rd generation of the $LoNO_x$[25,26] melter represents a melter with recuperative heat preheating without segmentation, with a low NO_X value due to the low air preheating temperatures of 600-700 °C. Most furnace designs are distinctly different from the recommended basic concept as recommended by Beerkens[27]. The re-circulation flows direct thus heating the batch blanket. Ehrig[28] describes this as an internal batch preheating, where the heat of the combustion gases is transferred onto the batch during batch-to-melt conversion in the first segment of the melter. By the direct preheating, the amount of dust increases because the glass dusts by the abrasion is additional to the filter dust, leading parallel to an increase of abrasion at the transport devices. Still, the consolidation of hot cullet and cold batch leads to dust development. The segmented melter was developed for pull rates about 3.5-3.8 t/m^2·d. At the non-container glass application, in the EU 25 the pull rate with 2.6 t/m^2·d is too low for an efficient application of this alternative melting concept. Nebel[29] shows the energy demand of 3700-3800 MJ per tons of glass for the $LoNO_x$ melter. These values can also be reached at [23], but requiring the building of a new furnace. Here, the focus is on an add-on technology in spite of a furnace rebuild with a new segmented melter.

The results of Heye Glas[30] on the energy savings, as well as to the CO_2 savings, have been verified according to the necessity of consistent energy data by Beerkens [14]. The data of energy and the efficiency is relative to an energy saving to 5.4 %. As conclusion for the oxygen-fuel firing melter with cryogenic oxygen generation, Pörtner[31] describes the necessity of modifications of all plants with filter modules to match the emissions of the melt, as well as the product related limits, based on a equivalent calculation. The energy in the flue gases from glass furnaces is often not explored. Heye Glas[32] has executed a project that demonstrates the efficient use of energy out of the hot flue gases (approx. 1450 °C). The flue gases pass through a steam boiler and the generated steam drives two turbines. Air blowers are connected to the shafts of these turbines and supply air to the hollow glass machines. The remaining heat is used for preheating broken glass material that is used as input raw material for the glass furnaces. The energy conservation reaches 746,000 Nm³ natural gas equivalent. Conclusively, the pay-back period of 6.2 years, upon a calculated 2.4 years with respect to the energy saving being three times higher with the use of the cullet preheater. To run the turbines completely under high vapour pressure, the temperature behind the turbines must be 225 °C. Herewith, the operation of multilevel turbines was suggested. The cullet can be preheated theoretically up to 110 °C (Steam inlet at 225 °C). This can be reached, if the steam

[24] http://www.quinn-group.com/Glass.html
[25] Pieper, H. Derzeitiger Erfahrungsstand beim Betrieb des Lonox-Melter. Mitteilung der HVG (1988) 3
[26] http://www.sorg.de/pdf/flex2_04.pdf
[27] Beerkens, R. Future industrial glass melting concepts. Proceedings of the International Confress on Glass, Vol. 1, Edingburgh, Scotland, 2001, 180-190
[28] Ehrig, R. Betriebserfahrungen mit $LoNO_X$-Meltern der ersten und zweiten Generation, HVG-Mitteilung Nr. 1917, 1997
[29] Nebel, R. Potential development for saving energy and reducing pollutant emissions. Proceedings of the International Novem Workshop Energy Efficiency in the Glass Industry, 200, Amsterdam, pp. 57-61
[30] Pörtner, D., Betriebs- und Emissionsverhalten von Glasschmelzöfen mit Brennstoff-Sauerstoff-Beheizung. Zwiesel, 2001
[31] Pörtner, D., Betriebs- und Emissionsverhalten von Glasschmelzöfen mit Brennstoff-Sauerstoff-Beheizung, HVG-Kolloquium, Frankfurt, 1999
[32] http:/ www.novem.nl/default.asp?documentId=10047 Demonstratieproject Novem Nr. 392117/7179. Hergebruik energie uit rookgasen van een oxyfuel glasoven, 2000

flow of the turbines of the last of the three cullet preheaters cools down in the cullet storage instead of in a separate condenser. Because the investment cost outbalanced the energy savings, the system was not realised although the potential upon electricity or steam generation is obvious.

Richards[33] demonstrates that even minute quantities of water in the batch would lead to sticking or backing phenomena. Cold batch is being gravity fed into the top of a heat exchanger tube. At lower level, the batch reaches 100 °C, at which water is vaporized. The escape of vapour is eliminated at hotter lower levels, when rising to colder levels above, the water recondenses. The water migration is also a topic at the indirect preheating system because the accumulated water causes the batch particle to stick together and hinders the flowability of the granular bulk solids. This can be prevented by recycling a part of the preheated bulk solids into cold bulk solids in a mixer on top of the heat exchanger. Werner[34] found a relation between the adhesive water and the particle size of the cullet while studying the segregation of the batch. The remaining water can be up to 10 % at fine cullet.

The addressed problem of dust development [20] is realized with an advantage at the indirect preheater by separation of the mass flow of the waste gases and the granular bulk solids. The rentability with regard the lower energy efficiency of the indirect preheating have to be taken into account as well as the need of additional installations, maintenance as well as dedusting devices of the direct preheating devices during operation.

8. SCIENTIFIC AND TECHNICAL OBJECTIVES IN SPECIFIC:

To optimise the indirect preheater for a successful second generation much earlier and ongoing experimental investigations are addressed to avoid the "infant mortalities" of the first generation. In order to reach the general objectives of the project, the following scientific and technical objectives have to be realised:

- Influence and technological limits on energy consumption by batch preheating, fuel and oxidant (oxygen versus air), cullet content and batch humidity,
- Output of technological changes on glass quality, melting behaviour, energy efficiency, emission and refractory wear,
- Heat and energy balance of new synergy approach (preheating and oxy-fuel firing),
- Batch Preheating: location near to the doghouse or integrated re-design of the doghouse as well as odour generation, due to organic fumes release, decrease of raw materials and prevention of sticking of material or stoppage of the batch flow
- Batch charging: Aspects of handling of cold batch (raw materials) and hot cullet or vice versa, increase of particulate emission of dry batch (dust) at furnace entrance
- Batch melting: Reaction kinetics, enhanced melting, shape of batch blanket
- Optimization or adoptions of construction, material, design, and maintenance, due to the higher outlet temperatures of waste gases at oxy-fuel combustion (corrosion phenomena, water content, aggressive batch and fuel components).

Some aspects, such as the water migration and the charging of the hot and cold bulk solids into the furnace, under analysis according to the preheated batch and cullet related melting behaviour, are given below.

---

[33] Richards, R.S. A batch preheater for energy efficiency. Ceram. Bull., Vol. 67. No. 11, 1988, p. 1802-1805
[34] Werner, A. J. Batch-Cullet Segregation Studies. Ceram. Eng. Sci. Proc. 8, [3-4], s. 217-221, 1987

9. WATER MIGRATION

The relevant physical chemical properties next to the mechanical properties influence the flow properties of a bulk solid and are relevant to that design of preheater. Next to the particle size distribution, the particle shape, the particle surface is as already discussed in [34] the moisture content important for the transport and heat transfer during preheating. The water forms arches between the grains or cullet (Here: tine, spherical, 1 % humidity), thus increasing adhesion forces. With decreasing particle sizes the bulk solids behave more cohesive. In a batch the granular movement as well as to the tendency of segregation is reduced by the humidity reduces. which improves the use of different grain sizes in batches.

Figure **4** shows the caking of the batch by reactions with water or CO_2 **from** the air, especially for materials being slightly soluble as e.g. soda ash, can lead to charging, melting, or conveying problems.

Figure **4** Caking at the switch of the preheater and the hopper and stoppage on top of the preheater in an indirect preheater due to long storage periods and water migration.

Primarily, the batch water reacts with the well-known impregnation effect, meaning the reallocation of the soda ash in the cold batch. The batch water enhances the early occurrence of batch reactions as the primarily melt formation, the fining, and the earlier release of CO_2 . In contrast to this indirect effect, there is a tremendous increase of the heat demand during the batch-to-melt conversion. The main part of the batch water is driven out at temperatures below 150 °C. This explains the course of the curves around 100 °C in Figure **5**. The specific energy consumption decreases drastically with batch temperature at the charging end of the furnace. As discussed in before, the potential batch preheat temperature is higher compared to the end port regenerative furnace. Thus more than 20 % can he saved.

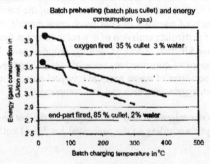

Figure **5** Effect of batch preheating on specific energy consumption end port fired regenerative and oxygen-fired container glass furnace for constant pull and constant fuel input. Grey circles are data points from existing furnaces. Excluding energy consumption oxygen generation [14].

Small amounts of water are kept in the system until approx. 1100 °C. Cullet reacts alone, and especially with the soda ash. very sensitively – and mostly negatively on the presence of batch water because of the delayed quartz dissolution. Also, the transition from the open (or granular bulk solid) to the closed pore (or reaction foam) stage and the duration of a latter phase are most important for redox control, color generation, and primary emission control and dependent on the water content in the batch.

The effect of cullet and batch water content[35] with regard to the preheated cullet and/or batch as well as the water release during the batch to melt conversion[36] has to be analysed in this study as a function of temperature with the microscopical phase transitions of the batch materials supported by the thermo chemical calculations.

The permeation of vapour through the granular bulk solids, for e.g. fine spherical cullets is observed in a vertical funnel. As a function of the temperature and the rate of the phase transitions the relevant flow properties e.g. unconfined yield strength, internal friction angle, wall friction angle, bulk density, time consolidation dependent on the consolidation stress are determined in a modified Jenicke shear tester. An ongoing study [45] observes the rheological behavior of the batch. Kuckelberg[37] studied the thermal conductivity and the transition of the heat radiation. Conradt[38] uses a modified ring shear tester of Jenike to measure heterogeneous bulk solids, as hatches. The modification allows the analysis of chemical reactions of bulks solids (water release, CO_2 release, formation of primary melts). Herewith, the shear cell can be held on individual temperatures up to 1200 °C, or heating-up with a constant rate. Almost no experimental observations are available on the batch viscosity. In
Figure 6, three phases a clearly distinguished. For the design of the preheater the first phase is seen as relevant. In phase I the warming up of all raw materials take place next to the evaporation of water, the energetically insignificant quartz transition, and the first decomposition step of dolomite occurring this phase.

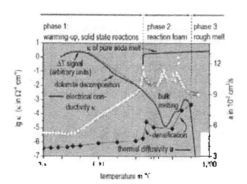

Figure 6. Electrical conductivity κ and thermal diffusivity a as determined [35] in 200 g tests on a commercial flint glass batch as a function of temperature.

[35] **Dubois, O., Conradt,** R. Experimental study on the effect of cullet and batch water content on the melting behaviour of flint and container glass batches. Glass Sci. Technol. 77 (2004), 137-148
[36] Conradt, R. Wasserfreisetzung während der Gemengereaktion.Fachausschuß III der HVG, Limburg, 10/2006
[37] Kuckelberg, D.. Becker, E., Conradt. R.: Experimentelle Bestimmung von Wärmeleitung und Strahlungswärmeübergang bei der Gemengeschmelze. 79. Glastechnische Tagung, Würzburg, 2005
[38] Grimm, M., Krzoska, J., Trebbels, R.. Conradt, R. Entwicklung einer Scherzelle für reaktive Schüttgüter bei erhöhten Temperaturen. 80. Glastechnische Tagung. Dresden, 2006

The batch essentially remains a granular bulk solid. The warming up phase shows [35] until 500 "C no considerably reactions, only the evaporation of water at 100 °C, between 500 and 800 °C solid state reaction take place without a high mass transfer rate. The results at the shear tester [38] show only small shear forces at this phase of warming up, slightly increasing with temperature. During the solid state reactions the maximum shear forces are measured. The course of the curve meets the behaviour of a consolidated sample. For the flowability the solid state reactions play a significant role, although there is almost none mass transfer in the batch nor influences the thermal properties of the batch. These observation will be transferred on the comparable reactions in the preheater.

Most preheater problems are a result of the way the bulk solid flows within the preheater. However, all modifications should be checked first on the basis of the shear tester result. With these results, it can be predicted which hopper inclination is required for mass flow. Here, the most important information is the flow profile, i.e. the way the bulk solid flows within the silo. With the mass transfer and heat transfer coefficients the layout of the preheater can be optimally estimated according to the funnel dimensions and the optimal flow behavior of the preheated granular bulk solids.

10. CHARGING OF THE PREHEATED CULLET AND/OR BATCH

The effect of preheated batch on batch-to-melt conversion will be studied with respect to the heat turnover and melting behavior. The lowering of the furnace and melting temperatures coherent with the replacing of pre-reactions of the batch/cullet into the preheater leads to the extension of the furnace campaign, and even more to the enhancement of the batch-to-melt conversion.

Investigations on the flow behaviour of bulk solids in the preheater itself have to be made next to observations as to the effect of the heat flux into the batch, the rate of batch-to-melt turnover, the internal heat sink term regarding the batch melting process with preheated batch. Trier's[39] approach to batch melting was primarily focused on the scale of phenomenological heat and mass transfer processes. He also introduced a paradigm to view the batch melting process as a temporal sequence of a warming-up, an initial, a stationary, and a final melting phase. Ungan and Viskanta[40] were among the first authors to take into account the scales ranging from grain-to-grain interaction to extended transport phenomena. Intensive research is done to characterize and quantify the boundary conditions and reaction phenomena of batch melting.

Until today, present studies have concentrated on the interaction between the furnace atmosphere in the combustion space and the batch in the basin of the glass tank, as a pull rate restricting mechanism, even more regarding the forced energy savings.

After the assessment of the raw materials for a good performance, the next step is high temperature tests in a semi technical scale (10 kg charges) to guarantee a trouble-free operation under aspects of bulk density, consolidation phenomena, dusting, mixing behavior. segregation, and the melting behavior.

---

[39] Trier, W. Glasschmelzöfen. Konstruktion und Betriebsverhalten.Kap. 6.4 Einlegen und Abschmelzen des Gemenges, 145-156. Springer Verlag, Berlin, 1984
[40] Unpan, A., Viskanta, R.: Melting behavior of continuously charged loose batch blankets in glass melting furnaces. Glastechn. Ber. 59 (1986), 279-291

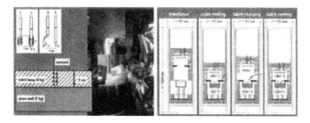

Figure 7 Observation furnace at semi technical scale (10 kg charges) for analysing the **melt**-ing behaviour *cf* preheated batch and for testing alternative charging methods **of** preheated granular **bulk** solids.

On top of a cullet melt preheated to **1400** °C and cold and hot bulk solids (cullet and/or batch at **400** °C) will be dropped and varied in terms of charging and bringing together, Figure 7. Variations are e.g. intensive mixing of cullet and rest of the batch, cold batch on top of hot cullet, vice **versa**, or as sandwich inlay. 'The dust development is measured, when combining the hot and cold mass flows of the granular bulk solids.

A better tank economy is related to the reactions for which the melting temperatures are in fact too high. This batch reaction can be brought forward under the release of batch gases coherent with the lowering of the partial pressures at the main melting pressures. Reducing of partial pressure means, at the same time, the reduction of the melting temperatures. With this, the energy savings are shown. Earlier melting and agglomeration of dust particulates of the batch and cullet reduces also the emissions. The melting behavior and **the** particulate emission at furnace entrance are evaluated **to** the charging strategy.

The objective is to determine the reactions in the melting batch locally and time dependent. Here, a thermal diffusivity sensor based on the run time of heat pulses through the batch are used, next to the direct observation during the tests. The so-called batch free time method is used to determine the melting rates function of the preheating temperature. The dissolution of sand grains in **a** batch is determined by grain-to-grain interactions with other raw materials **as** well as with alkaline or earth alkaline oxide rich primary formed melts **as** shown by Sam-kam[41] and used by Verheijen[42]. Own investigations, Figure **8** demonstrate the dissolution of coarse sand grains as well as natural quartz **stones**, supported by microscopy.

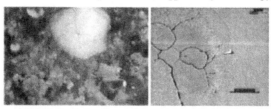

Figure 8 Diffusion controlled dissolution through a reaction product layer **of** a natural **quartz** stone. Cristobalite dendrites and trydimite laths **in** a glassy alkali-alkaline earth enriched transition phase next to the **quartz core in the glass**[43]

[41] Sam Kham, N. Entwicklung von Methoden zur Charakterisierung einschmelzender Glasgemenge (Development of Methods for the Characterization of Melting Batches), Dissertation RWTH Aachen, (2005)
[42] Verheijen. Oscar. Thermal behavior of glass forming batches, Dissertation, TU Eindhoven. 2003, ISBN 90-386-2555-3
[43] Glüsing, A. Dissolution kinetics of oxide and metallic impurities in cullet during glass melting. Auflösüngsverhalten von keramischen und metallischen Verunreinigungen aus Glasscherben während der Glasschmelze, Dissertation RWTH Aachen (2003) ISBN 3-68130-394-9

Thermal conductometry and measurements of the chemical heat demand are, together with the parameter of reaction kinetics, the basis for modelling of batch reactions during melting by analyzing the depth of the heat radiation through the upper layer of the batch. The extension of the melt surface covered by batch influences the heat transfer into the glass bath as a function of the melting velocity and the viscosity of the batch. In particular, the thermal behavior of glass forming batches [42], foaming[44], and rheological properties of the batch and rough melt'' are addressed in this study.

11. SUMMARY

However, such a demonstration does not come without substantial risks. It is in this regard that public support is needed for this program in order to accelerate the technology's final development and demonstration. This support enables demonstrations to be made at two scales and will provide the industry with the pathway necessary to implement this technology at other **glass** plants in all branches of the glass industry. **As** the technology is proven and process benefits quantified, it is anticipated that the application to conventional air fuel furnaces will follow. Several other glass industry market segment companies have expressed their interest in this technology demonstration's review. The turnover of the European glass industry was 8.2 bN in the year 2004[46]. Therefore, this branch is a leading industry in the European Community. In order to strengthen the competitiveness and their market position in the world, the creation and launch of new products and transfer of concepts for creativity. conductivity, versatility a working, learning and living world is a key activity of the innovation process.

[41] Läimböck, P.R. Foaming of Glass Melts, Dissertation TU Eindhoven (1998) ISBN 90-386-0518-8
[45] Grygarova, L., Deubner. J., Trebbels, R., Krozska, J. Conradt, R. Development of methods of comprehensive screening of batch rheology. 80. Glastechnische Tagung, Dresden, 2006
[46] www.hvglas.de

MELTING AND REFRACTORIES

MELTING AND REFINING CONDITIONS IN AN ALL ELECTRIC MELTER FOR BOROSILICATE GLASS

Matthias Lindig
Nikolaus Sorg GmbH & Co.KG
Lohr am Main, Germany

SUMMARY
The all-electric glass melter represents a melting technology suitable for lower pull, high liquidus temperature glasses and or glasses with high volatilization rates. The borosilicate **glass** type is one of those glass compositions typically molten in an all-electric melter. Advantages of the application are the cold batch cover of the melt suppressing most of the evaporation and the direct heat transfer from the electrodes into the melt. The general problems of the all electric melting technology are the fining conditions and the degassing of the melt. In case of the borosilicate type glass the low solubility of the sulfur requires careful design of the furnace itself and almost constant process conditions. Investigations **on** sulfur solubility in borosilicate glasses have been performed and are well described in the literature. Based on these results it can be shown that the electrode configuration immersed from top through the batch results in better degassing conditions compared to the configuration with bottom electrodes. This approach is in accordance with a number of patents taken out by different inventors addressing the sensitive flow and degassing conditions of this type of melter. CFD Simulation model tests results are given for the different design solutions. The results confirm the beneficial conditions achieved by the top electrode configuration.

GENERAL ASPECTS REGARDING ALL ELECTRIC MELTING
The basics of all electric glass melting have been invented very early when the continuous glass melting and production came **up.** First patents were granted in the beginning of 1900. The principle is the electrical resistivity of the glass melt as a function of the composition and temperature. The benefit of the technology is apparent. The energy is transferred directly into the melt. The superstructure, the combustion chamber, being an additional source of heat **loss** due to outside wall radiation as well as the waste gases still carrying a remarkable portion of energy to the stack are ceased. The cost for electric power being secondary energy generated from fossil energy or water power plants is the limiting factor for broader application of that technology. The technology is also limited in its application in terms of total pull and from a certain point of view in achievable quality and flexibility.

The limitation in pull and furnace size is due to batch coverage and the need of almost uniform energy input. A belt charger length will be limited at about 6-7m. Due to this the footprint for larger furnaces has been changed already from square to rectangle (larger in width). The distance between the electrodes for larger furnaces will require bottom electrodes instead of side or top electrodes. In general the operation requires a more or less even batch cover with a constant thickness in order to achieve stable temperature conditions. Larger variation in pull or cullet addition will strongly affect these batch layer conditions.

The material flow in a fossil fired melting unit is oriented horizontal. The raw materials are charged from one end. The batch is heated up and generates a flux. During the conversion more or less gases are released due to decomposition of the raw materials. The gases are releases on the hot surface in the charging area and the area with free surface. The re-fined melt with an acceptable low number of seeds is leaving the melter through the throat.

In case of the all electric melter the melting process is oriented vertically which is obviously less convenient for the re-fining, degassing conditions. The melter basin is covered with batch which should reduce heat losses from glass to ambient and also keep the melt just below the batch blanket hot enough in order to guarantee the temperature regime at this level, which is necessary for releasing all the seeds. The location of the electrodes in the melter basin should provide the best conditions for melting and to ensuring the necessary flexibility for pull changes.

Sorg® has developed his own SORG® VSM® melter concept in the early 70th. Physical modelling was applied in order to study the flow patterns of that special design concept.
Later, the concept was investigated and improved later by means of mathematical simulation. Because that method is based on a more accurate data base of applied materials and glass properties it becomes the preferred method for SORG® regarding any design studies carried out for a large number of projects.

The principle of the SORG VSM® is to put the heat source in the area very close to the batch area. Side electrodes result in severe side wall wear and are not the preferred solution. The top electrodes provide a concentration of the heat input there where it is highly requested. The position of the heat source results also in a strong current close to the surface while the lower part of the furnace is characterized by slower, less intense flow patterns. The refined glass flowing towards the throat should get the opportunity to reabsorb the remaining blisters. The mixing of refined glass melt with seedy glass melt directly coming from the surface should be avoided.

The second typical feature of the SORG® VSM® melter is the rotating crown.
The rotating crown should provide a sealed charging area, avoid heat losses and ensure very even batch cover. Finally the design is less space consuming. There is possible access to the top of the furnace from all sides.

CHARACTERISTICS OF BOROSILICATE MELTING AND RE-FINING
As mentioned above the melting conditions in an all electric melter differ significantly from a fossil fired melting unit. The batch and re-fining agent addition has to be adjusted in order to reduce the quantity of gases being released during the conversion of solid batch to liquid phase. Haft et al. (1) have described in very much detail the special conditions concerning melting of borosilicate type of glasses and more over the neutral glasses.

In Table I is given a number of typical glass compositions in wt% for borosilicate type glasses with a thermal expansion coefficient between $33 \cdot 10^{-7}$ to $42 \cdot 10^{-7}$/K.

Table I: evaluation of borosilicate type glass compositions with different expansion

	Duran	Pyrex	III	IV	V	VI
SiO_2	80,48	80,8	80.29	80,17	80,15	73.51
B_2O_3	12,61	12	11,21	12.85	12,06	9,82
Al_2O_3	2.42	2,2	2.23	2,38	2,15	6.74
Na_2O	3.4	4.3	6.31	4,04	5.56	6.7
K_2O	0,57		0,7	0,09	0,015	2,53
CaO	0,02		0,05	0,26		0,65
Fe_2O_3	0,03			0,045	0.015	
F						0,08
Cl	0,06		**0,2**			

The borosilicate glasses generally show a very low solubility of sulfur oxides. The solubility of halogenides as well is depending strongly on the composition.

The SO_3- solubility if different glass types are in the following ranges:

	SO_3 solubility in wt% at 1500°C
Soda-Lime	0,05 - 0,25
E-glass	0,01 - 0,015
borosilicate type	< 0,001

The solubility of the gases (chemical or physically dissolved) has a strong influence on the re-fining conditions and the sensitiveness regarding reboil. For some glasses reboil may take place when the glass melt is heated up again or when the redox state changes behind the melting process. Clausen et al. (2) have investigated the thermodynamics of the Sulfur in borosilicate glasses. They found that the gaseous SO_2 is released at higher temperatures with an increased content of alkalines. Koepsel has investigated the solubility and vaporization of halogenides (3). In his laboratory tests be proved that the solubility of the halogenides is increased with increasing basicity. These basic investigation results explain why the neutral glasses or borosilicate glasses with increased amounts of alkalines are more difficult to re-fine with the traditional re-fining agent Sodium Chloride. Re-fining is used in the sense of primary fining for release the fining gases and consequently bubble growth and ascension in the melt. The Sodium Chloride only evaporates in these melts at higher temperature. In case of all electric melters usually only a very small quantity of fining agent is used in the batch in comparison to fossil fired melting process thus the volume of gases to be removed will be decreased.

In case of any unstable conditions in the all electric melter we have to face a blister and or seed upset in the glass downstream at the production. Analyzing the blister content we find in these blisters in case of borosilicate glasses typically almost SO_2 sometimes with a deposit of Chloride. This is due to the low SO_2 solubility of these glasses.

RE-FINING **PROBLEMS** AND **TYPICAL COUNTERMEASURES**

Typically the blister upset in an all electric melting process occurs after pull change, changes in cullet addition, batch cover instabilities or strong changes in energy input. Reasons could also be a temperature drop in the furnace or strong reheating the channel downstream the melter unit or blisters are caused by changes in batch composition. Generally everything which has an impact on temperature and flow distribution field in the melter as well as on the chemistry could be a cause for a blister upset. It has to be kept in mind that the furnace design itself will also have a strong impact on the sensitiveness of the melting process against all the mentioned parameters.

A number or countermeasures are usually applied which are more or less addressing the basic problem. The countermeasures are summarized in table II.

Table II: typical countermeasures and validation

countermeasures	validation
raise NaCl addition	not necessarily beneficial, can have an influence on the sulfur dissolution and degassing process as well as on the reboil
add Niter (Na or K-Nitrate)	presumably no impact, Niter decomposes quite early; might help to stabilize the ferric oxidation state
add oil	usually only added in cold top furnaces for dusting reduction, might have an impact on total redox state near below batch
add fluorine	should act as a flux in the batch layer, low viscous phase for easier degasing conditions throuhout the batch layer
reduce cullets	avoid early glassy phase generated by the cullets; higher temperature below batch, stronger insulation and better degasing conditions
raise temperature	only possible to a given limit in order to maintain a certain batch layer thickness also limited by the acceptable temperature nearby the bump charger if there is one
raw material check	change to raw materials with less degassing potential i e from Alumina hydrate to Feldspar or from Borax to Dehybor

Most of the counter-measures address the conditions in the boundary layer between batch and melt. Only very nearby the batch cover the blisters have a chance to leave the melt. If the temperature regime in this area is too low or if the downward current is too fast the degassing takes place at a time when the glass is almost on the way to escape through the throat. The buoyancy force is far too small to help the blister to ascend to the melter surface.

Even smaller quantities of sulfur in the batch mostly introduced as impurities result in a blister problem in case that the temperature and current **flow** conditions are affected. According to the measurements of Clausen (2) the solubility of the sulfur is reduced with increased temperature. The interaction between the sulfur and another polyvalent ion can again worsening the conditions. Depending on temperature conditions (see reference 2) the sulfite can react with the iron according to the basic equation:

$$2FeO + SO_4^{2-} \Leftrightarrow Fe_2O_3 + SO_2 + O^2$$

This might have convinced some experts to reduce as much as possible the iron content in the sand (below 0,15wt%). If the temperature regime below the batch is reduced the Iron oxide can oxidize the Sulfate (reverse reaction) and help to stay dissolved in the melt, but leading to reboil at higher temperature.

GENERAL DESIGN SOLUTION FOR RE-FINING IMPROVEMENTS
There are a number of patents granted giving a clear picture of how the industry tried to battle the re-fining problems in an all electric melter.

There exists more than one example for adding to an all electric melter a second chamber which is used for re-fining either fossil fired or also electric heated (5,6,7). The proposals have in common the open surface in the second chamber which should enable to release the blisters.

Another approach is to prevent the surface glass melt going fast forward downwards near the side walls and leave the melter unit through the throat. For instance a duct made of Molybdenum should be placed bottom side connected with the throat ending in the center of the melter unit (7,8)

An interesting approach is given in a patent recently published (9). According to the invention the throat channel is extended up to the bottom center of the furnace. The front side of the channel immersed into the bottom is covered with a shielding. Similar to the tube (see ref. 7,8) the glass is only being released from the center of the furnace to the exit. Another approach given in that patent is to reduce the cold top melter surface by placing a shielding immersed into the glass front side above the melter exit.

The given selection of the published inventions shows that the control of the downward flow in the all electric melter has a major influence on the blister quality of the glass. The time and temperature history and the temperature field inside the melter have to be influenced by either melter design features or the energy input into glass which can be manipulated the electrode position.

MODELLING STUDIES ON ALL ELECTRIC MELTER TYPE
Sorg has carried out a large number of mathematical modeling tests in order to investigate the different all electric melter design approaches and in order to improve the own technology. The flow patterns in the melter can be evaluated in a real furnace as well as in a physical or mathematical model by so called tracer experiments. A small quantity of a detectable element (in case of the model only a large number of particles) is charged for a short time period. The time distribution of the response detected at the melter outlet is characteristic regarding the individual mixing behavior. More detailed information about tracer tests is given in (10). The model was tested by Sorg in the beginning in order to investigate how precise the result can predict the real furnace conditions. A C-glass furnace with a pull of 49t/d and and $25m^2$ melting area was modeled. A tracer study was carried out at the real furnace. The results were compared with a particle tracing received from the model post processing. The residence time distribution was in good agreement with the real fumace tracer results.

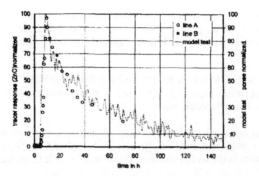

Fig.1 : Real furnace and model tracer result comparison

The influence of the melter basin depth on the residence time was investigated by model case studies. The melter depth has a significant influence on the residence time distribution. The model results were again compared with real furnace data from Sorg melter and others with lower basin level.

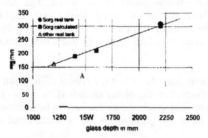

Fig.2 : Influence of the all electric melter basin depth on the **minimum** residence time.

The furnace in Fig? with a glass depth of about 1400mm and very low minimum residence time of 120min is an all electric melter with bottom electrodes

The modeling study shows a clear relationship between basin depth and residence time. The results of real furnace fit into these results. The real furnace residence time in figure 2 significant below the regression line is related to a furnace with bottom electrodes. All the other results are representing furnaces with top electrodes. This indicates clearly the main concern and problem regarding bottom electrodes. The position results in a strong current flow at the bottom section resulting in an early release of the melt through the throat, the furnace exit. Lower minimum residence time and the lack of glass conditioning processing step are responsible for low blister quality performance. Furnace depth and position of electrodes seems to be crucial for ensuring glass quality in case of all electric melting.

The Sorg" VSM" furnace design was compared to a typical all electric melter with bottom electrodes a similar pull and melter size. The furnace data are given in Table III.

Table III: Modelling comparison all electric melter with top and bottom electrodes

glass		Sorg borosilicate	bottom elec. brosilicate
melting area	m²	19,6	16
glass depth	m	2,1	1,3
throat		straight	submerged
pull	t/d	36	27
spec pull	t/m²d	1,84	1.69
electrodes		12 top	12 bottom
energy input	kW	1640	1040
cullet	%	25	25
capacity	t	79	48
mean resid time	tl	52	36

Particle tracing calculations were carried out in both models. The results are given in the following figure 3. The 2 furnaces are displayed in cross section with the traces of a selection of the first. the fastest particles. The first response in case of bottom electrodes is more than one hour earlier compared to the case of the melter with top electrodes and a deeper basin

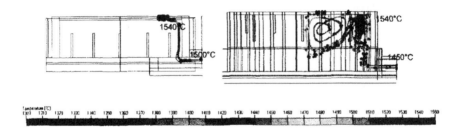

Fig.3: Results of particle tracer test in mathematical modelling - critical trajectories of the melt in the two all electric melters, left case with bottom electrodes, the right one with top electrodes

In case of the top electrodes the hottest area and the strongest current flow in the melter is nearby the surface. There is a strong circulation **below** the batch. Even the particles on the fastest way out are re-circulating near the top. Below the electrode level the glass is slowly descending and is getting cooler. These conditions should be more convenient for the degassing of the melt.

With bottom electrode the risk of obtaining a fast forward downward current flow near the wall towards the throat is quite high.

CONCLUSIONS
The use of the all electric melter technology for borosilicate glass is more sensitive concerning glass quality (residence times. blister number) compared to fossil fired melting units and requires adjustments in raw material selection and re-fining agent addition. The cold top temperature regime makes it difficult especially for borosilicate type of glass to release the batch gases.

Significant reduction in re-fining agent addition as well as raw materials with low degassing potential is requested. The positioning of the electrodes from bottom to top may have an advantage in order to keep the surface layer just underneath the batch blanket as hot as possible. The optimum batch cover thickness in an narrow range, the temperature regime and uniform energy input requires significantly limiting the variation in pull and cullet addition. Due to the low sulfur (SO_2/SO_3) dissolution capability of the melt, the degassing has to be complete to avoid reboil or too late SO_2 release.

REFERENCES

1) **R.Haft**, H.J.Linz; Investigations into all-electric melting of neutral glass with improved reboil behavior; Glasstechn.Ber.Glass Sci.Technol. **67 (1994),4, 93- 98**

(2) O.Clausen, R.Seuwen, D.Koepsel, B.Unterberger; Glass Sci.Technol. **(2002)75,169-175**

(3) D.Koepsel; Solubility and Vaporization of Halogenides; Glastechn.Ber.Glass Sci.Technol. **73 C2 (2000) 43- 50**

(4) German patent DE **43 13 217 C1, 9/1994;** Inventor Jenaer Schmelztechnik Jodeit GmbH
Process device and processing for all electric melting of neutral glass

(5) US patent **5,194,081, 3/1993;** Inventor R.E.Trevelyan, P.J.Whitfield (Pilkington)
Glass melting process

(6) US patent **4,693,740, 9/1987;** Inventor R.Noiret, M.Zortea (Saint Gobain)
Process and device for melting fining and homogenizing glass

(7) US patent **4,029,887,611977;** Inventor P.Spremulli (Corning Glass Works)
Electrically heated outlet system

(8) **US** patent, **4,737,966, 4/1988;** Inventor R.Palmquist (Coming Glass Works)
Electric melter for high electrical resistivity glass materials

(9) World patent WO **01177034 A2, 1012001;** K.D.Duch, F.Karetta (Schott AG)
Device for producing a glass melt

(10) Muschick, Mysenberg; Round Robin for glass tank models: Glastechn.Ber.Glass Sci.Technol. **71(1998)6,153-156**

RECENT DEVELOPMENTS IN SUBMERGED COMBUSTION MELTING

David Rue, John Wagner, and Grigory Aronchik
Gas Technology Institute

ABSTRACT

The Gas Technology Institute (GTI) is developing submerged combustion melting (SCM) as the melting and homogenization stage of the Next Generation Melting System (NGMS). GTI is leading a consortium including Corning Incorporated. Johns Manville, Owens Corning, PPG Industries Inc., Schott North America, and the Glass Manufacturing Industry Council in this effort. The project team appreciates financial support from the U.S. Department of Energy, the U.S. gas industry. and the glass company members of the project consortium. The team has melted a range of industrial glass batches in a lab-scale SCM, has carried out physical and CFD modeling of the SCM process, and has designed and nearly finished fabrication of a 1 ton/h continuous SCM unit for further testing.

BACKGROUND

The submerged combustion melter is a bubbling type furnace capable of producing glass and cementitious melts from a number of materials (geologic rocks, sand, limestone, slag, ash, waste solids, etc.). In the submerged combustion melter, illustrated in Figure 1, fuel and oxidant are fired directly into the bath of material being melted from burners attached to the bottom of the melt chamber. High-temperature bubbling combustion inside the melt creates complex gas-liquid interaction and a large heat transfer surface. This significantly intensifies the heat exchange between the products of combustion and the processed material while lowering the average combustion temperature. The intense mixing of the melt increases the speed of melting. promotes reactant contact and chemical reaction rates. and improves the homogeneity of the glass melt product. Another positive feature of the melter is its ability to handle a relatively non-homogeneous batch material. The size. physical structure, and especially the homogeneity of the batch feed do not require strict control. Batch components can be charged either premixed or separately, continuously or in portions.

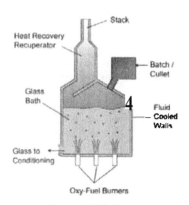

Figure 1. SCM Concept

A critical condition for successful SCM operation is stable, controlled combustion of the fuel within the melt. Simply supplying a combustible mixture of fuel and oxidant into the melt which is at a temperature significantly above the fuel's ignition temperature is not sufficient to create stable combustion. Numerous experiments conducted on different submerged combustion furnaces with different melts have shown that the simple injection of a combustible mixture into a melt (for example. through tuyeres) does not ensure stable combustion. Cold channels are formed that lead to explosive combustion and excessive melt agitation. A physical model for the ignition of a combustible mixture within a melt as well as its mathematical description show that for the majority of melt conditions **that** may occur in practice, the ignition of a combustible mixture injected into the melt **as** a stream starts at a significant distance from the injection point. This, in turn, leads to the formation of cold channels of frozen melt, and explosive combustion. To avoid this type of combustion, the system must be designed to minimize the ignition distance. This can be achieved in three ways: 1) by stabilization of the flame at the point of injection using special stabilizing devices, 2) by splitting the fuel-oxidant mixture into smaller jets, and/or 3) by preheating the fuel/oxidant **mixture**[1,2]

In the melt bath, heat exchange between the high-temperature products of combustion and the batch particles primarily occurs through the melt. This process occurs in two steps: 1) heat exchange between the products of combustion and the melt, and 2) heat exchange between the melt and the batch particles. Studies have shown that the first step is controlling when producing melt for the manufacture of mineral wool. For glass melts and when silica is >40%) of the batch, the second step (dissolution of SiO_2) is controlling[3].

PHYSICAL AND CFD MODELING

In an effort to determine the optimum melter cross-sectional area and burner pattern, a series of cold flow modeling tests were conducted with glycerin. Air jets were placed below, shooting upward into, a plexiglass tank of glycerin simulating molten E glass. Dimensions, jet flows. **and** jet velocities were chosen in dimensionless groups to allow for accurate simulation of an SCM unit[4]. Setting the Reynolds and Grashof numbers equal between the dimensions of the planned pilot-scale melter and the glycerin model, the viscosity-temperature characteristic of E glass were matched'. This approach is unable to simultaneously account for heat transfer effects, but these were handled in later CFD modeling.

A series of tests **were** conducted in the cold flow model with rectangular, circular, and octagonal chambers along with many jet configurations. Tests were conducted by establishing stable conditions of glycerin feed and discharge. bed level, and air jet rates followed by an injection of an aliquot of dye. **A** colorimeter was used to monitor glycerin discharge samples to determine initial dye discharge and rate of dye mixing Figure 2 shows the physical modeling unit and the results of testing. Several general conclusions were reached. First. symmetrical burner patterns were **best** to prevent dye bypassing to the discharge. Second. the best residence time control is achieved with sufficient melt depth for convective flow paths to form around the jets. **And,** finally. circular, and octagonal cross sections are desirable to avoid stagnant zones in the comers **of** a rectangular bath.

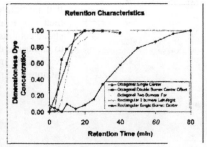

Figure 2. Physical Modeling Unit **and Dye Retention Curves**

CFD modeling was carried out with Fluent modified to handle three-phase flow and varying temporal events including combustion and convective mixing. **CFD** calculations **confirmed the** physical modeling results and showed that the most uniform temperature-viscosity profile along the length of a rectangular or octagonal SCM unit is achieved by firing burners at different rates. This is apparent when the first burner **(or** row of burners) must provide most of the heat **of** melting, while the later burners (or burner rows) provides mixing and added residence time.

CFD results **are shown** in Figure 3 for an octagonal pilot-scale melter with sex burners in 3 **rows** of 2 **and** a feeding section set off from the first row of burners. These calculations confirmed that nearly half **(45%)** of the heat duty should be provided by the first row of burners. while the middle burner row provides less than a quarter of the heat duty.

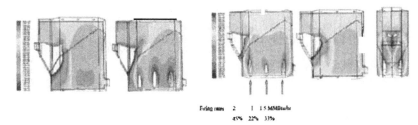

Figure 3. Calculated Pilot-Scale **SCM** Temperature Profile on a Line Through **the** Burners for (a) Uniform Firing, and (b) Non-Uniform Firing

An SCM **unit** is a high-intensity melter with short average residence times of only a few hours. The efficiency of SCM increases dramatically **as** the residence time decreases. There is, however, a limit to the maximum **SCM** pull rate. even if sufficient thermal energy is introduced into the melt. This limit is set by the minimum residence time of all feed particles. with silica being to slowest dissolving batch constituent. **For** typical E **glass** batch particle size, approximately **1** hour **is** needed to fully dissolve all silica **grains**[6]. **SCM** produces a distribution of residence times with **a** minimum time set by melter shape, burner pattern, and pull rate. Figure **4** shows calculated residence time curves for E glass batch in the pilot-scale SCM melter. Figure **5** presents the range **of** expected SCM thermal efficiencies as a function of pull rate as the melter size is increased. At pull rates of 120 to 140 ton/day, SCM efficiency. with no heat recovery is calculated to be at or above **50%**. equivalent with the best current **oxy-gas** melters.

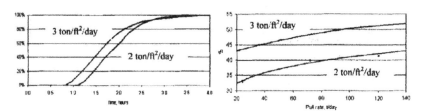

Figure **4.** Calculated **SCM** Flow Element Residence Times

Figure **5.** Calculated SCM Thermal Efficiency vs. **Pull** Rate

LAB-SCALE SCM TESTS

Before building a continuous feed and discharge pilot-scale melter, a range of commercial glass compositions were melted from batch in a semi-continuous lab-scale **SCM** unit. This unit is fired with two oxy-gas burners, each with a maximum firing rate of 1.5 MMBtu/h. A batch feeding system with calibrated weigh screws and an exhaust baghouse are included. The maximum pull rate for this small 21 **x** 31 in. rectangular melter, shown in Figure **6**. is approximately **400** lb/h.

The glass compositions melted from batch included soda-lime (CertianTeed batch), E glass (Owens Corning and **Johns** Manville batch), and LCD glass (Corning batch). Batch was found to be fully melted and homogeneous, but there **were** significant bubbles in all product glasses. Melt bed temperature and platinum tap temperature were increased going from soda-lime. to E glass, to **LCD** glass. All were handled withing the firing range of the lab-scale melter.

The project team also melted scrap fiberglass including Owens Corning Advantex scrap and Johns Manville wool insulation scrap. In both scrap fiberglass tests all resin was removed from

Figure 6 Lb-Scale SCM

the glass. No carbon particles were observed in the product glass, and carbon content was below the detectable limit

PILOT-SCALE SCM UNIT

The project team has analyzed both melter performance and product glass data from the lab-scale SCM unit and designed a 1 ton/h continuous feed and discharge SCM unit. The unit design, shown in Figure 7, is based on a rectangular cross section, with corners eliminated, to create an octagon with no 90° corners. The melt bath internal dimensions are approximately 35 x 55 inches. Melt depth will he variable from 1 to 4 feet. There are six burners. each capable of firing up to 1.5 MMBtu/h. and each independently controlled for both firing rate and oxygen to fuel ratio. A new, larger platinum discharge pipe has been designed and fabricated. To allow for flexibility, two feeders, one From the roof, and one on the wall above the melt bath, have been included. Exhaust gas is from two side ports designed large enough to drop exhaust gas velocity to minimize particulate carryover. The melter is constructed from a series of water cooled panels that will have a layer of 1.5 in. of castable refractory installed and held in place by steel anchors. Melt rate will be high enough at up to 1 ton/h that a discharge trough was designed. This trough has a moving stainless steel belt that carries and cools glass under a shallow- pool of water and then lifts the cooled product glass for discharge into waiting hoppers. Glass samples for analysis will be collected by hand as the product melt stream from the platinum discharge pipe falls toward the water surface in the discharge trough. For shutdown, a second discharge port has been included that will allow the operators to quickly empty all molten glass from the melter. This will allow faster turn around between tests.

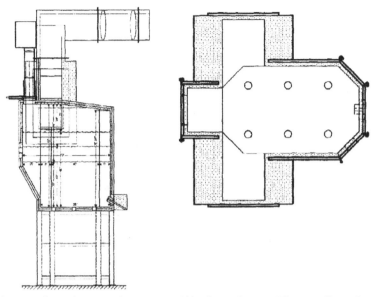

Figure 7   Pilot-Scale SCM Unit Showing (a) Side View. and (h) Top View **With** Burner Locations

The melter panels have been fabricated, assembled for check for fit of all the angles, and water leak tested. All panels have passed leak testing. They are now being disassembled in order to have anchors and refractory installed. Once refractory is in place, the melter will be assembled on the stand in the test cell. The assembled melter is shown in Figure 8. without the eshaust gas ducts included. **A** total of six new burners have been fabricated. The burner tips have been plasma coated with a zirconia coating to evaluate extended life with the refractory and to prevent any possibility of trace metal contamination **of** the product **glass** during melting. The burners have all been fabricated following the same design as the burners used in the lab-scale SCM unit. New burner designs are under evaluation at GTI **on** a test stand, but they are not yet ready for testing **on** a working melter. The **new burner** designs are very promising and show a very stable. short flame that is expected to release all heat within the melt while not being anchored so deep in the burner that damage is done to the burner body.

Figure 7. Pilot-Scale SCM Unit

The pilot-scale SCM unit will be fully instrumented with all data collected digitally using a LabView data acquisition system. A special platinum thermowell with three internal thermocouples will be inserted up into the melt through the melter floor in order to read melt hatch temperatures. All other instrumentation will be routine. but a number of sample ports have been installed in the melter walls so additional measurements can be made at later time, if desired, during testing. A summary of the pilot-scale SCM unit instrumentation package is shown in Figure 8.

Gas and oxygen flow	Into all 6 burners
Gas and oxygen pressure	Into each burner
Water flow	Into each melter panel
Differential pressure	At melter flue exhaust
Melt temperature	Internal thermowell
Temperatures	Gas, oxygen, water, tap, exhaust gas, refractory
Nuclear level gauge	Average bed height
Digital cameras	Melt surface and tap
Voltage, amps	Tap transformer control
Weights	Batch feed rate

Figure 8   Pilot-Scale SCM Unit Instrumentation Package.

After pilot-scale SCM unit assembly and shakedown. a series of continuous tests will be conducted with the same glass compositions (soda-lime. E glass, LCD glass, and scrap fiberglass) that were melted in the lab-scale SCM unit. These tests will evaluate melt quality. thermal efficiency. melt depth, firing patterns. residence time distributions, pull rates, melt bath temperatures, and other independent variables.

Future work, also funded by the U.S. Department of Energy. the New York State Energy Research and Development Authority, the gas industry, Owens Coming, and Johns Manville. will focus on evaluating several methods for rapidly removed bubbles from the SCM product glass. Methods to be evaluated include steam refining, sonic refining, and short residence time retiring. Other methods will be modeled but not tested on the pilot scale melter.

With completion of the pilot-scale melter and the first test series, the facility will be available for help the project team take the steps needed to move NGMS and SCM toward commercial use. The needed steps will include industrialization of the pilot-scale unit. the addition of needed instrumentation and controls, and demonstration testing in a glass factory under industrial conditions.   This path will provide a sound engineering based introduction of the Next Generation Melting System into the glass industry.

REFFRENCES

1. Olabin, V.. Pioro, L.S.. "Igniting Fuel in the Melt – Tedriya I Practika Szhiganya Gasa, Vol. VI.L., NedRa, 187-92, 1975.
2. 'Olabin, V.. Krivtsun, N.N,, -'Method of Determining the Final Batch Depth in Melters", Khimicheskaya Technologya, 6, 23-26, 1979.

3. Olabin, V. , Maximuk, A., Rue, D., Kunc, W., "Development of Submerged Combustion Technology for Producing Mineral Melts", Int. Gas Research Conf. (IGRC), San Diego, CA, Nov. 1998.
4. Tolstov, V.A., "Physical Modeling of the Glass Melt Hydrodynamics in a Highly Efficient Melting Furnace for Container Glass", Glass and Ceramics, vol. 57,2000.
5. Harper, Ch.A., 2001. Handbook of Ceramics, Glasses, and Diamonds, McGraw-Hill, New York, NY, USA, pp. 5.34–5.35.
6. Nemec, L. , "Engineering and Chemistry of the Glass-Melting Process". Pure Appld. Chem., 11, pp. 21 19-29,2002,

NEW SOLUTIONS FOR CHECKERS WORKING UNDER OXIDIZING AND REDUCING CONDITIONS

G. Heilemann', B. Schmalenbach', T. Weichert[1], S. Postrach', A. Lynker[2], and G. Gelbmann[3]

[1] RHI Glas GmbH, Abraham-Lincoln-Strasse 1,65189 Wiesbaden/Germany
[2] RHI Refractories. Didier Werke AG, Didierstrasse 30-32,35460 Staufenberg/Germany
[3] RHI Refractories Technology Center, Magnesitstrasse 2,8700 Leoben/Austria

INTRODUCTION

During the last years there are an increasing number of demands on the refractory materials for the checkers of Soda-Lime-Glass furnaces. Especially the need for fulfilling the new environmental regulations as well as the demand for energy savings due to the fact of highly increased energy costs have changed the requirements. So the working conditions of a glass furnace are different and even the use of special procedures for waste gas treatment e.g. the 3R-method are influencing the waste gas conditions and therefore the attack on the regenerator checker refractories has changed. In addition to oxidizing conditions, more and more reducing atmospheres can be found and, even more dangerous, changing conditions are observed over the life time of a furnace. The regenerator checker refractory materials that one can get on the today's market often do not fulfil this demand and fail under the current conditions. Therefore the refractory industry is required to develop new products. By the new materials: Rubinal PZ, Rubinal MA R and Rubinal ESP for the first time materials are available which are eligible as a "switch-hitter" for oxidizing as well as reducing conditions with good results. This paper recapitulates nowadays conditions for checker materials, up to date regenerator lining recommendations and new material developments.

COMPOSITION OF THE WASTE GAS

The waste gas of a Soda-Lime-Glass melting furnace mainly consist of combustion products of the fossil fuels (oil or gas) such as CO, CO_2, HCl, HF, SO_X, NO_X, O_2 and N_2. A reducing atmosphere is signified per definition by a CO-content of >1000 vol.-ppm. But also several volatization products of the glass melt (particularly $NaOH$) and non flammable or incompletely incinerated residues of the fuels (e.g. Vanadium, Ash etc.) can be found. Dusty batch ingredients are carried with the waste gas from the batch blanket in the furnace by the turbulent flames.

ATTACK ON THE CHECKER WORK UNDER OXIDIZING CONDITIONS

Top courses
The main attack of this checker area is due to the carry over and consequently the erosive and corrosive attack on the bricks. If oil is used as a fuel also an attack due to the V_2O_5 from the oil has to be expected. Affected are only the top 2-4 courses of the checker work. Below this level the percentage of these corrosive ingredients is low and the abrasion is negligible.

Due to the deposit of batch carry-over on the top courses formation of new mineral phases within the refractory material happens. SiO_2 from carry-over can cause the formation of Forsterite in contact with Periclase (MgO). This is combined with a volume growth which deteriorates the brick structure. As a protection against this reaction, Magnesia-Zircon-bricks are used. In these bricks the Periclase grain is protected by a fringe of Forsterite and Zirconia.

In fused cast AZS the glassy phase of the bricks is attacked and the bricks also fail due to the so called thermal overstress as the bricks soften and deform while used at to high temperatures. Therefore these bricks are not any longer used in this area.

In case of a combined attack of SiO_2 from sand and CaO, which originates from recycled filter dusts and tine cullets, even Magnesia-Zircon bricks are destroyed. The CaO attacks the Forsterite in the bonding phase forms Monticellite (CMS) and the bricks loose their structure and the high temperature resistance declines. Solutions which work under such conditions are ceramically bonded Alumina **or** fused cast Alumina bricks.

An additional scenario which is found at the top courses is the smelting of slags which drip down in the hot areas and solidify in the cooler areas. This again affects the operation conditions negatively due to blockings of the channels.

Middle courses
In the temperature range down to 1100°C usually there is no severe chemical attack on the bricks if the furnace is a natural gas heated furnace. In the case oil is used as fuel the attack of V_2O_5 has to be taken in account. V_2O_5 attacks all CaO containing materials like the group of Dicalcium-Silikate(C2S)-bonded or so called pure MgO bricks. Under oxidizing conditions eutectic Ca-Vanadate and under slightly reducing conditions volatile Ca-Vanadite is formed. By these processes the CaO is deprived from the bonding phase of Magnesia bricks and the *CIS* ratio is driven from stabile C_2S to lower melting phases Monticellite (1440°C) and Merwinite **(1570°C)**. All these reactions cause the destruction of the bricks or the deformation due to the high mechanical load respectively. For oil-fired furnaces bricks with low CaO content like Magnesia-Zircon bricks has to be chosen.

Temperature range < 1 **100°C** (condensation zone)
The waste gas of most of the furnaces contains SO_2 as a volatile product of the refining agent (Sulfates) or as an component of the fired oil. Part of SO_2 reacts with oxygen in the waste gas to SO_3. The also present NaOH reacts with SO_3 by forming Na_2SO_4 and H_2O. Below 1100°C the Sulfates condense and infiltrate into the brick structure. Excessive SO_3 can react with the brick components and lead to destruction of the bricks. Especially Periclase reacts to Sodium-Magnesium-Sulfates. In the same manner the CaO in the brick can be attacked by Sodium-Sulfates. The newly formed Sulfates condense **or** solidify within the brick structure and subsequently the elasticity and the thermal shock resistance of the bricks is reduced. Due to emerging cracks the brick can decay which is known as Sulfate bursting.

Since many years Magnesia-Zircon bricks have proven themselves in this area. The bonding phase of these bricks exist of Forsterite and Zirconia. The Periclase grains within the bricks are also protected by a fringe of Forsterite and Zirconia. Both parts are not attacked by Sodium-Sulfates or SO_3.

Today it is usual to equip the complete checkerwork down to the rider arches with this Mag-Zircon brick type. By doing so the danger of corrosion and thermal overstress of low grade materials like Fire Clay and Sillimanite is avoided especially considering the possible need for thermal cleaning. Table I summarizes the most important reactions at oxidizing conditions.

Table 1 : Attack and reactions under oxidizing conditions

op courses and middle courses		
1. SiO_2 –attack		
$MgO + SiO_2$	$\rightarrow$	Forsterite (M_2S)
Dicalciumsilicate(C_2S)+ SiO_2+MgO	$\rightarrow$	Monticellite (CMS), Mrrwinite(C_1MS_2) und M_2S
2. CaO-attack:		
$M_2S + CaO$	$\rightarrow$	CMS , C_3MS_2
3. V_2O_5-attack		
C_2S, CMS, C_3MS_2 + V_2O_5	$\rightarrow$	CA-Vanadates and CA-Vanadites(vol)
CaO-depletion followed by $C_2S \rightarrow C_3MS_2 \rightarrow CMS \rightarrow M_2S$		
condensation zone (700 – 1100°C)		
1. Sulfate formation :		
$2 SO_2 + O_2$	$\rightarrow 2$ **sol**	
$SO_3 + 2 NaOH$	$\rightarrow Na_2SO_4 + H_2O$	
2. direct attack of SO_3:		
$MgO + SO_3$	$\rightarrow MgSO_4$ (<900°C)	
$CaO + SO_3$	$\rightarrow CaSO_4$	
3. attack of Na_2SO_4:		
$MgO + Na_2SO_4$	$\rightarrow$ Na-Mg-Sulfates	
No attack on M_2S		

ATTACK ON THE CHECKER WORK AT REDUCING WASTE GAS CONDITIONS

For the top and middle courses the statements for oxidizing conditions apply accordingly. But there are differences at the condensation zone.

Temperature range <1 100°C
A reducing atmosphere influences the materials in the regenerator in 2 different ways:
➤ destruction of refractories with a Fe_2O_3 content of >1%
➤ Increased attack of Alkali-compounds on the refractories

Influence on refractories with high Fe_2O_3-content:
At reducing atmosphere Fe_2O_3 is reduced to FeO while the hot blasting cycle (waste gas passes the checker pack). During the cold blasting cycle (the air passes the checker pack) this reaction is reversed. The continuous change of the oxidation state and the related volume changes finally lead to a loss of the single brick structure and in sum to a collapse of the total checker pack. Due to these stresses bricks for reducing applications preferably contain Fe_2O_3 below 1%.

Increased attack of Alkali-compounds
As earlier described the waste gas of most of the furnaces contains SO_2 as a volatile product of the refining agent (Sulfates) or as an component of the fired oil. Part of SO_2 reacts during cooling with oxygen in the waste gas to SO_3. As earlier described SO_3 reacts at temperatures below 1100°C with NaOH to Sulfates and H2O. At reducing atmosphere the waste gas contains no oxygen and therefore the SO_2 is not oxidised. Therefore also the forming of Sulfates is strongly limited. The Alkali remain unbonded in the waste gas and due to their chemical reactivity the Alkali react with acid Oxides (especially SiO_2). Due to this fact all refractories containing >0.8% SiO_2 are strongly attacked under reducing conditions as the following examples show:

Fire Clay, Sillimanite and Mullite: Glassy phases are formed which lowers the softening point of the materials by up to **150°C.** Image 1 shows a Fire clay brick after a simulating test as described later in this paper within the temperature range 800-1**100°C** under reducing conditions. On the contrary this softening can not be observed under oxidizing conditions. Under reducing atmosphere the now limited refractoriness under load has **to** be considered while using Sillimanite, Fire clay or Mullite. Especially by using Mullite the formation of Nepheline and as a result fatal spalling is observed.

Image 1: Fire clay under reducing atmosphere

Fused cast AZS: At the brick surface the formation of Nepheline takes place due to reaction of the material with the Alkali-compounds in the waste gas. Due to the difference in thermal expansion between Edge and core of the brick, spalling is observed. Once the dense surface is destroyed the porous core of these bricks is exposed and can easily be attacked and destroyed. Image 2 shows such a brick after running the simulation test performed in the same way as the one for the Fire clay brick.

Image 2: Fused cast AZS under reducing atmosphere

Forsteritic bonded **Magnesia and Magnesia-Zircon bricks:**
The chemical **attack** is mainly on the forsteritic bonding phase which is destroyed. **The** bricks **loose** their structure and **get** crumbly. Image 3 shows such a brick after **running a** simulation **test performed** in the same way **as** the one for the Fire clay brick.

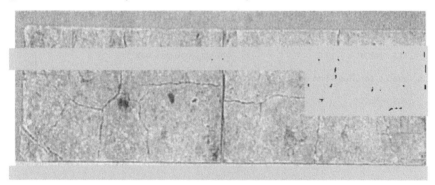

Image 3: Magnesia-Zircon under reducing atmosphere

Because of the described reactions of the Alkali-compounds with the bricks under reducing **atmospheres** only bricks with low SiO_2-**contents**(<1%) can be used with **good** success. This demand is fulfilled for example by the modern C_2S-bonded Magnesia bricks or pure MgO-bricks with high portions of direct bonding **as** it is reached by high firing.

For **a** better understanding Table **11** once again **summarize a the most** important attacks and reactions which can be **observed** while running **a** regenerator at reducing conditions.

Table II: Attack and reaction under reducing conditions.

top courses and middle courses		
1. SiO2 –attack		
MgO + SiO$_2$	→	Forsterite (M$_2$S)
Dicalciumsilicate(C$_2$S)+ SiO$_2$+MgO	→	Monticellite (CMS), Merwinite (C$_3$MS$_2$) und M$_2$S
2. CaO-attack:		
M$_2$S + CaO	→	CMS , C$_3$MS$_2$
3. V2O5-attack		
C$_2$S, CMS. C$_3$MS$_2$ t V$_2$O$_5$	→	CA-Vanadates and CA-Vanadites(vol)
CaO-depletion followed by C$_2$S→C$_3$MS$_2$→CMS→M$_2$S		
condensation zone (700 – 1100°C)		
1. only strongly reduced Sulfate formation therefore no attack of Sulfates		
2. attack of NaOH on Forsterite:		
M$_2$S + NaOH		→ Na-Mg-Silicates + H$_2$O
3. attack of NaOH on Silimanite, Fire Clay. Mullite, AZS:		
A$_x$S$_x$ + NaOH		→ Nepheline t Al$_2$O$_3$ (ΔV=6-36%)
4. attack of NaOH on Alumina:		
Al$_2$O$_3$ (a-Alumina) + NaOH		→ β-Alumina (ΔV=28%)

Due to the gathered knowledge of the above mentioned reactions of the waste gas with the refractory materials the following lining recommendations have evolved.

LINING RECOMMENDATIONS

Oxidizing conditions

1) **Top courses (2-4):**
Magnesia-Zircon e.g. **Rubinal VZ, Narmag VZ**
Fused cast β''''-Alumina: e.g. **ER 5312RX**
Ceramically bonded fused Alumina: e.g. Durital K99extra

2) Middle courses:
C₂S-bonded Magnesia (gas fired)   e.g. Anker DGI, Rubinal VS, Super Nar-
(partly direct bonded)   mag B, Narmag 98B
Magnesia-Zircon (oil fired)   e.g. Rubinal VZ, Narmag VZ
Fused cast AZS   e.g. ER 1682RX

3) layers < 1100°C:
Magnesia-Zircon   e.g. Rubinal EZ, Narmag EZ
Fused cast AZS   e.g. ER 1682RX

Reducing conditions

1) Top courses:
Magnesia-Zircon   e.g. Rubinal VZ, Narmag VZ
Fused cast β'''-Alumina: e.g. ER 5312RX
Ceramically bonded fused Alumina: e.g. Durital K99extra

2) Middle courses:
C₂S-bonded Magnesia (gas fired):   e.g. Anker DG1, Rubinal VS, Super Nar-
(partly direct bonded)   mag B, Narmag 98B
Magnesia-Zircon (oil fired)   e.g. Rubinal VZ, Narmag VZ
Fused cast AZS   e.g. ER 1682RX

3) Layers < 1100°C:
C₂S-bonded Magnesia low iron   e.g. Anker DGI, Rubinal VS, Super Nar-
(partly direct bonded)   mag B, Narmag 98B

As it can be seen from the above recommendations no lining problems exist for the top and middle courses of checkers under oxidizing conditions as well as under reducing conditions. But for the temperature range below 1100°C no material exists which works successfully under both or changing conditions. This was the demand and the driving force for the new developments and even needs more developments for the future.

REQUIRED PROPERTIES FOR THE ACTUAL DEVELOPMENTS

While considering the experiences and the demands of the industry a refractory material has to be developed fulfilling the following needs:

- ➢ Chemical and mechanical resistance under reducing and oxidizing conditions.
  - ➢ Low iron content
  - ➢ Low SiO_2
- ➢ Good thermal conductivity
- ➢ Good heat transfer
- ➢ No contact reactions with the conventional checker materials
- ➢ Adequate cold crushing strength
- ➢ Thermal stress resistant
- ➢ Creep resistance up to 1300°C
- ➢ Moderate price compared with the already used materials
- ➢ Harmless at disposal

CHECKER WORK SIMULATING TEST

Already more than 15 years ago RHI developed special checker work simulating tests. With this test. refractory materials can be checked under different atmospheres regarding its qualification as checker material.

Image 4 shows the assembly of the test. Before the test specimens of 50*40*150mm are put in a tubular furnace in a checker setting. The actual test grid is protected against extensive heat losses by a second blind grid. The test grid can be exposed to several additives likes batch carry-over, Alkali-compounds, Sulfates etc. which are injected directly into the flame. As a burner an Oxygen-Propane-burner is used.

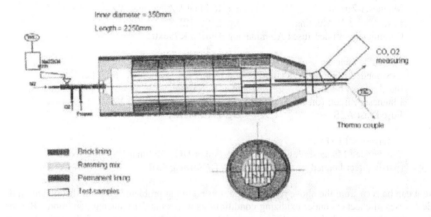

Image 4: Regenerator Simulation Test

The test conditions for testing the qualification of materials for a use in the condensation zone(T<1100°C) of a regenerator is set as follows: Preheating of the furnace to around 800°C. Afterwards cyclic operation of the furnace within a temperature range between 800 and **1100°C**. The test lasts 96 cycles of 1 hour each (30 minutes heating up. 30 minutes cooling down). This simulates the hot blasting and cold blasting such as it occurs in the checkerwork of a regenerator of a glass furnace. The test specimen are analyzed subsequently microscopically as well as macroscopically. Depending on the demand oxidizing or reducing atmospheres as well as even changing conditions can be simulated by the test. For testing the materials regarding their performance under other temperatures respective temperature setting can be used for this test.

NEW PRODUCT DEVELOPMENTS

After taking everything into account, only a few oxidic new refractory compounds were found which are stable under the described conditions. Apart from ZrO_2 especially MA-Spinel is interesting due to its good resistance against both Sulfates and Alkalines.

For this reason a refractory quality based on a pure fused MA-Spinel is an obvious solution and was developed. The resulting quality is named Rubinal MA R. In image 5 shows the microstructure of this new quality.

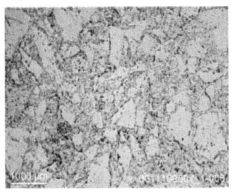

Image 5: microstructure Rubinal MA R

Rubinal MA R is characterized by its homogenous grain distribution combined with a very strong sintering and a good embedding of the coarse grains. Both are essential aspects for a high corrosion resistance.

A completely different approach led to the development of the second quality named Rubinal PZ. It is based on imbedded, coarse MgO-grains in a ZrO_2 matrix. The microstructure in this material can be seen in image 6. It shows that two main components are present.

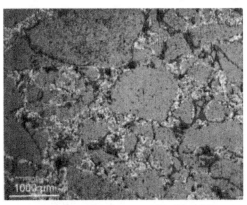

Image 6: microstructure of Rubinal PZ

Both components have good resistance against Alkali-ompounds as can be found under reducing conditions. For the use under oxidizing conditions the idea is that the corrosion resistant ZrO_2 matrix protects the MgO-grain against the attack of the Sulfates. In order to verify this assumption the material was tested in a simulation test. **As** a result only a very small attack of the MgO was noticed in the border area **of** the specimen (**s.** image 7).

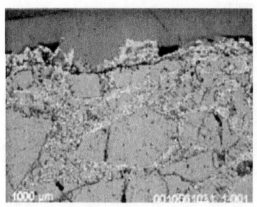

Image 7: microstructure of Rubinal PZ after simulating test (800-1100°C).

The amount of attack is comparable to the attack on the already existing Magnesia-Zircon qualities which have excellent resistance under oxidizing conditions.
Thus Rubinal PZ is a excellent quality to serve under reducing as well **as** oxidizing conditions.

Furthermore these test results verify that this concept to design a "corrosion resistant matrix for a MgO grain" works successful. Rased an these results, recently another brick grade was developed.

In variation to Rubinal PZ, in Rubinal ESP the MgO-grain is imbedded in a Spinel-matrix. The remarkahle detail on this development **is** the especially strong bonding between the matrix and the MgO-grains which is due to specific inventions. Image 8 shows the special microstructure of this grade

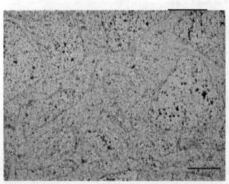

Image **8:** microstructure Rubinal ESP

Due to this strong bonding the matrix protects the bricks structure in the same way as in Rubinal PZ. As a result the corrosions resistance of this brick is high under oxidizing conditions and should also be very good for a use under oxidizing conditions as a result from the first test which were done.

	Rubinal MA R	Rubinal PZ	Rubinal ESP
Bulk density in g/cm³	2,95	3,34	2,95
Open porosity in Vol-%	16,5	14,5	15
Cold Crushing strength in MPa	70	80	65
TC acc. Klasse @ 500°C	4,0	3,4	
TC @ 750°C	3,2	3,3	
TC @ 1000°C	2,8	3,2	

Table IV: chemical composition of the new products

Oxide in weight-%	Rubinal MA R	Rubinal PZ	Rubinal ESP
MgO	33,8	74,0	73,0
Al₂O₃	62,7	-	23,8
SiO₂	1,0	0,4	1,5
CaO	0,4	0,7	0,9
Fe₂O₃	0,5	0,4	0,5
ZrO₂	-	24,0	-

By developing these three qualities, new superior solutions for checker applications are available. All three materials are especially designed for a use under oxidizing and reducing or even changing conditions. First industrial installations are under evaluation for these materials. More laboratory testing combined with even more field tests and industrial trials will be done in the near future. As a matter of fact these new developments allows us also in the future to recommend always the best refractory solutions which becomes due to the described changes and influences more and more tailor-made issue.

COLLECTION OF LITERATURE REGARDING THIS SUBJECT:

[1] Beanspruchung der Gitterung in den Regeneratoren von Kalk-Natron-Silicatglaswannen durch Einwirkung von Abgas-, Gemenge und Glaskomponenten, F. Gebhardt et. al., HVG-Kurse 1983

[2] Beitrag zum Einsatz feuerfester Steine in den Regenerativkammern von Flachglaswannen, F. Gebhardt, presentation in Unterausschuss „Feuerfeste Baustoffe" of DGG, 1977

[3] Feuerfeste Steine im Gitterwerk von Natronkalkglaswannen – Zustellung und Verschleiß-, T. Weichert, presentation in Ostberlin 1988

[4] Natron-Kalk-Glaswannen Gittermaterialien von Regenerativ-Kammern Verschleißmechanismen und Zustellungsempfehlungen, B. Schmalenbach et. al., Didier Information 1987. Wiesbaden

[5] Mechanism of attack on upper course checker bricks in glass tank regenerators, G. Evans et. al., Glass Technology Vol. 13 No. 4 August 1972

[6] Corrosion resistance of binding and matrix phases in basic checker bricks for the middle regions of glass furnace regenerators, C. Wohrmeyer et. al., Glastechn. Berichte 63 Nr. 5, Frankfurt 1990, S 1 18 ff.

[7] Ersatz und Weiterentwicklung von Magnesia-Zirkon-Steinen in der Glasindustrie, T. Weichert et. al., proceedings of XXXVI. International Colloquium on Refractories, Aachen 1993, S.76 ff

[8] Structure-property relationships in Magnesium-Aluminate Spinel bonded Magnesia Refractories for Glass furnace regenerators, S. Plint et. al., proceedings of XXXVI. International Colloquium on Refractories, Aachen 1993, S.79 ff

[9] Application of Spinel checker brick for glass furnace regenerators, M. Ogata et. al., proceedings of XXXVI. International Colloquium on Refractories, Aachen 1993. S.84 ff

[10] Eine neue Magnesiasteingeneration für gemengestaubbeanspruchte Zustellungsbereiche in Regeneratoren von Glasschmelzöfen, G. Mögling et. al., proceedings of XXXVI. International Colloquium on Refractories, Aachen 1993, S.95 ff

[11] Building on experience with Magnesia-Zircon checkers, B. Schmalenbach et. al., Glass international 6/97

[12] The impact of the reduction of Nox to nitrogen, cullet and filterdust recycling on regenerator refractories, K. Wieland, prodeedings of ATIV conference 1999

[13] New solutions for regenerators of glass melting furnaces, B. Schmalenbach et. al., RHI Bulletin 1 2003, Leoben

[14] Einsatz von keramisch gebundenen Korund-Werkstoffen im Oberbau, Gewölbe und Regenerator von Glasschmelzwannen. C. Mulch, presentation on 46. International Colloquium on Refractories. Aachen 2003

[15] New solutions for regenerators of glass melting furnaces, B. Schmalenbach et. al., proceedings of 46. International Colloquium on Refractories, Aachen 2003, S.25 ff

[16] New solutions for regenerators of glass melting furnaces, B. Schmalenbach et. al., presentation at the 7th ESG Conference on Glass science and technology, Athens 2004

[17] Konstruktive Mcrkmale von Warmeriickgewinnungsanlagen. proceedings of HVG-Fortbildungskurse 2005, S. 165 ff

ER 2001 SLX : VERY LOW EXUDATION AZS PRODUCT FOR GLASS FURNACE SIJPERSTRUCTURE

M. Gaubil, I. Cabodi, C. Morand, and B. Escaravage
Saint-Gobain CREE. Cavaillon. France

Exudation of AZS superstructure refractories (due to temperature and aggressive running conditions reducing atmosphere) can have an impact on the glass quality by enhancing superstructure run down. This phenomenon could involve glass defect formation such as vitreous enriched alumina defects with or without secondary zirconia crystals.

Figure 1 : **AZS** superstructureglass defect

In order to prevent superstructure glass defect we developed a new AZS product with a very low exudation level at high temperature.

This new fused *cast* material product, belong to the chemical composition field of AZS (Alumina Zirconia Silica) ternary system as illustrated in Table I.

Table I: Chemistry & microstructure of **AZS** products

%	ZrO_2	Al_2O_3	SiO_2	Na_2O	Zirconia	Corundum	Vitreous phase
ER 1681	32.5	51.2	15.0	1.3	32	47	21
ER 1851	36.0	53.1	10.0	0.9	35	51	14
ER2001Slx	17.0	68.3	13.0	1.7	16	63	21

This composition belongs to the primary field of corundum crystal formation and it is characteriseby the microstructure shown in Figure 2.

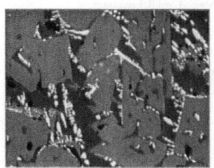

Figure 2: ER2001Slx microstructure

This microstructure is composed by corundum crystals, free zirconia crystals and an alumino silicate glassy phase.

We used to evaluate the exudation of refractory products two main exudation test. The first one, called "Louisville test", is also used as quality control by our plants, and consists in measuring the percentage of volume increase of a disc shape sample after a test of 16h at 1500°C.

The second one, called "bar test". consists in measuring the percentage of volume increase on a bar shape sample after two cycles from room temperature to **1500°C**. with a 4h plateau.

The results of the tests indicated for the new **AZS** fused cast material values lower than 0.5% in our two tests. This good property is due to a combination of different parameters such as
- Low quantity of impurities(id ER1681 / ER1851), its
- Limited amount of vitreous phase (21%, id ER1681).
- Good oxidation of the liquid bath.

and mainly to a microstructure effect associated to this particular chemical composition and solidification path.

Table II: Exudation values of **AZS** products

Exudation %	Louisville	Bar (1st cycle)	Bar (2nd cycle, cumulated)
ER1681	1,5	2.7	5.4
ER1851	0.7	2.6	3.4
ER2001Slx	< 0.5	< 0.5	< 0.5

The bar test sample after test (figure 3) and Louisville high temperature picture (figure **4)** test illustrate the very low level of exudation of this new AZS product

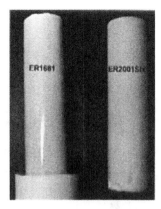

Figure 3: Bar test: ER1681 vs ER2001Slx

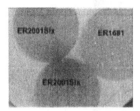

Figure 4: Louisville test : ERI 681 vs ER2001Slx
(high-temperature picture)

Some exudation tests were also performed at very high temperature (until 1675°C), under video observation. They show that the exudation of ER2001Slx remains also very low with these very strong solicitations (Figure 5).

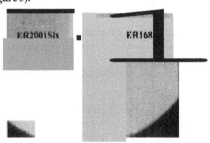

Figure 5: Picture taken during a bar test under video at 1600°C.

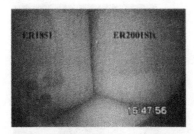

Figure 6: Picture taken after test under video at 1675°C

At very high temperature. the very low exudation of ER2001Slx induces the preservation of its original microstructure without creating porosity as in conventional AZS;.

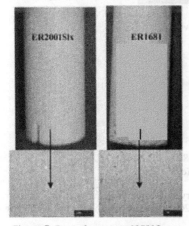

Figure 7: Bars after test at **1650°C**

This behaviour allows increasing corrosion resistance of this fused cast **AZS** product in superstructure running conditions (carry overs and alkali vapours attack)

By the way, this product **ha.** also other good properties for application in glass furnaces superstructure.

Indeed, ER 2001 SLX has **an** excellent behaviour in vapours phase corrosion, both in lab test and industrial furnace test, and against carry over. Penetration of soda seems especially limited by its particular microstructure (vitreous phase weakly interconnected, interlocked microstructure), **so** its transformation is not deep This is illustrated by the result of an industrial **test** in sodalime glass furnace

ER 2001 SLX: Very Low Exudation AZS Product for Glass Furnace Superstructure

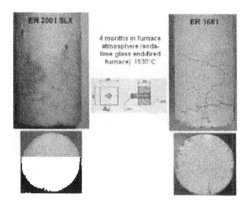

Distance from high temp surface Products	0 cm	2 cm	5 cm
ER 2001 SLX	**2.65** [+56%]	**2.40** [+41%]	**1.a4** [+8%]
ER1681	**2.94** [+126%]	**2.74** [+111%]	**2.47** [+98%]

% Na2O

[% enrichment in Na2O]

Moreover, ER 2001 SLX is situated between ER 1681 and ER1851 in term of resistance under load & creep. Especially, its refractoriness under load of 0.2 MPa is at 1700°C.

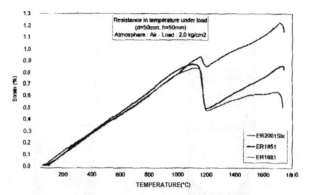

Figure 9: Refractoriness under load

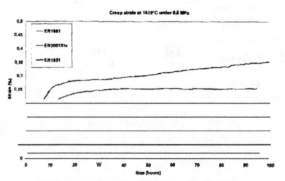

Figure 9: Creep test at 1620°C under 0.5 MPa

Finally this AZS has also an interesting dilatation curve (low shrinkage *at* zirconia transformation, good property for opening joints).

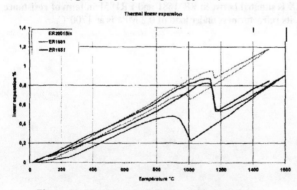

Figure 10: Linear expansion curve with temperature

Its thermal conductivity is between the ER1681 and the Jargal M (alpha − beta alumina product).

THERMAL CONDUCTIVITY OF AZS PRODUCTS

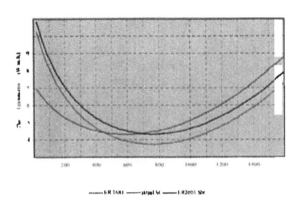

This new fused cast **AZS** product named ER 2001 SLX has been introduced in the SEFPro material offer since 2006.

Author Index

Author Index

Printed and bound by CPI Group (UK) Ltd, Croydon, CR0 4YY

16/04/2025

14658442-0009